University Scholar Conference: Engineering and Innovation (USC 2022)

IOP Conference Series: Materials Science and Engineering
Volume 1279

Kathmandu, Nepal
17 – 18 May 2022

ISBN: 978-1-7138-7142-2
ISSN: 1757-8981

Printed from e-media with permission by:

Curran Associates, Inc.
57 Morehouse Lane
Red Hook, NY 12571

Some format issues inherent in the e-media version may also appear in this print version.

Institute of Physics
Dirac House, Temple Back
Bristol BS1 6BE UK

Phone: 44 1 17 929 7481
Fax: 44 1 17 920 0979

techtracking@iop.org

Additional copies of this publication are available from:

Curran Associates, Inc.
57 Morehouse Lane
Red Hook, NY 12571 USA
Phone: 845-758-0400
Fax: 845-758-2633
Email: curran@proceedings.com
Web: www.proceedings.com

TABLE OF CONTENTS

IOP Conf. Series: Materials Science and Engineering 1279 (2023) 011001 doi:10.1088/1757-899X/1279/1/011001

Preface

The University Scholar Conference: Engineering and Innovation (USC) is a conference organized by the Association of Mechanical Engineering (AMES) at Kathmandu University (KU). The first edition, USC 21, was successfully completed in virtual mode, and the second edition, USC 22, was held on May 17-18, 2022, in a hybrid mode. USC 22 was hosted by Kathmandu University, co-hosted by the Himalayan University Consortium (HUC-ICIMOD), and supported by the Department of Mechanical Engineering - KU, the University Grant Commission (UGC), the Nepal Technology Innovation Center (NTIC), and the Green Hydrogen Lab (GHL). The inauguration ceremony of the two-day conference event was held at ICIMOD head office with a keynote session delivered by Dr. David Molden, Director General of ICIMOD, and the closing ceremony was organized at Kathmandu University.

The University Scholar Conference 2022 provided an excellent platform for graduate and undergraduate students to share their cutting-edge developments and state-of-the-art research work in the fields of science and engineering. One interesting fact about this conference is that it was completely organized by a student club under the supervision of faculty. Thus, the USC conference is a student-level conference *organized by university students for university students*. The conference was attended by a mix of academia, researchers, industrialists, professors, students, and policymakers. The conference committee highly encouraged the participation of graduate and undergraduate students engaged in engineering and science education within Nepal and abroad.

With the tagline "*Women Empowerment in Engineering*," four thematic areas were defined for USC 22: i) Automotive Systems, ii) Materials and Design for Transformation, iii) Digitalization and Emerging Technologies, and iv) Advances in Energy Systems. A total of 103 abstracts were received at the initial phase, out of which 48 papers were selected for presentation in the conference. Participants from 10 different universities from 5 different countries presented their research works in 12 plenary sessions during the two conference days. In addition to the parallel plenary sessions, a poster presentation was also conducted for funded projects. Based on the quality of the presented work, four best paper awards were declared for each thematic area, and a gift hamper was presented to the winners. The 48 papers presented at the conference were considered for further publication in the *IOP Conference Series: Materials Science and Engineering*. After a detailed peer-review process, a total of 10 papers were selected for final publication in this volume.

IOP Conf. Series: Materials Science and Engineering 1279 (2023) 011001 doi:10.1088/1757-899X/1279/1/011001

Overall, the University Scholar Conference: Engineering and Innovation (USC 22) was a success and provided a valuable platform for university scholars to share their research and engage with other professionals in the field of science and engineering. We hope that this conference will continue to be organized and provide opportunities for students to showcase their work and enhance their academic careers. We would like to express our sincere gratitude to the members of the Scientific Committee and Advisory Committee who made great efforts to review and revise the conference papers, ensuring the quality and scientific level of the papers. We hope that the proceedings will promote similar research practices at the undergraduate level.

Surendra Sujakhu, PhD

Editor-in-Chief and Coordinator, University Scholar Conference 2022

Faculty Coordinator, Association of Mechanical Engineering Students, Nepal

Assistant Professor, Department of Mechanical Engineering, Kathmandu University, Nepal

Prayush Jung Karki

Editor, University Scholar Conference 2022

Department of Mechanical Engineering, Kathmandu University, Nepal

IOP Conf. Series: Materials Science and Engineering 1279 (2023) 011001 doi:10.1088/1757-899X/1279/1/011001

Conference Chair

Dr. Bim Prasad Shrestha	Professor, Department of Mechancial Engineering, Kathmandu University, Dhulikhel, Nepal

Organizing Committee

Dr. Surendra Sujakhu	Conference Coordinator	Kathmandu University, Nepal
Mr.Pawan Sapkota	Conference Secretary	Kathmandu University, Nepal
Mr.Bivek Baral	Technical Management Chair	Kathmandu University, Nepal
Mr.Grishma Khatiwada	Technical Management Chair	Kathmandu University, Nepal
Ms.Kusum Khatiwada	Publicity Chair	Kathmandu University, Nepal
Mr.Avishek SUbedi	Publicity Chair	Kathmandu University, Nepal
Mr.Prayush Jung Karki	Publication Chair	Kathmandu University, Nepal
Ms.Anuapama Gaihre	Publication Chair	Kathmandu University, Nepal
Mr.Prajwal R. Biswokarma	Registration Chair	Kathmandu University, Nepal
Ms.Isha Acharya	Registration Chair	Kathmandu University, Nepal
Mr.Saras Mainali	Registration Chair	Kathmandu University, Nepal
Mr.Sagar Niroula	Web Service Chair	Kathmandu University, Nepal
Mr.Sugam Aryal	Web Service Chair	Kathmandu University, Nepal
Mr.Milish Dhungel	Corporate and Finance Chair	Kathmandu University, Nepal
Ms.Shrija Shrestha	Corporate and Finance Chair	Kathmandu University, Nepal
Mr.Ashmat Chhetri	Multimedia Chair	Kathmandu University, Nepal
Ms.Strena Shrestha	Multimedia Chair	Kathmandu University, Nepal
Mr. Aashish Dawadi	President, AMES	Kathmandu University, Nepal
Mr. Himanshu Giri	Secretary, AMES	Kathmandu University, Nepal
Mr. Mausam Bhurtel	Executive, AMES	Kathmandu University, Nepal
Mr. Sangam Bhandari	Executive, AMES	Kathmandu University, Nepal

Advisory Committee

Dr. Bhola Thapa	Vice-chancellor, Professor, Kathmandu University, Nepal
Dr. Bivek Baral	Professor, Kathmandu University, Nepal
Dr. Hari Prasad Neopane	Professor, Kathmandu University, Nepal
Dr. Manish Pokharel	Professor, Kathmandu University, Nepal
Dr. Janardhan Lamichane,	Professor, Kathmandu University, Nepal
Dr. Bhupendra Bimal Chhetri	Professor, Kathmandu University, Nepal
Dr. Daniel Tuladhar	Professor, Kathmandu University, Nepal

IOP Conf. Series: Materials Science and Engineering 1279 (2023) 011001 doi:10.1088/1757-899X/1279/1/011001

Dr. Rajendra Joshi	Professor, Kathmandu University, Nepal
Dr. Sunil Prasad Lohani	Associate Professor, Kathmandu University, Nepal
Dr. Krishna P. Shrestha	Associate Professor, Kathmandu University, Nepal
Dr. Biraj Singh Thapa	Associate Professor, Kathmandu University, Nepal
Dr. Binay K.C.	Associate Professor, Kathmandu University, Nepal
Dr. Shyam Sundar Khadka	Associate Professor, Kathmandu University, Nepal
Dr. Subash Ghimire	Associate Professor, Kathmandu University, Nepal
Dr. Bal Krishna Bal	Associate Professor, Kathmandu University, Nepal
Dr. Shailendra Kumar Jha	Assistant Professor, Kathmandu University, Nepal
Dr. Khagendra Acharya	Assistant Professor, Kathmandu University, Nepal
Dr. Bibek Uprety	Assistant Professor, Kathmandu University, Nepal
Dr. Sailesh Chitrakar	Assistant Professor, Kathmandu University, Nepal
Dr. Ranjit Shrestha	Assistant Professor, Kathmandu University, Nepal
Mr. Pratisthit Lal Shrestha	Assistant Professor, Kathmandu University, Nepal
Mr. Malesh Shah	Assistant Professor, Kathmandu University, Nepal
Mr. Bijendra Shrestha	Assistant Professor, Kathmandu University, Nepal
Mr. Hari Dhakal	Assistant Professor, Kathmandu University, Nepal
Mr. Niranjan Bastakoti	Assistant Professor, Kathmandu University, Nepal
Mr. Chiranjeevi Mahat	Assistant Professor, Kathmandu University, Nepal
Mr. Gokarna Paudel	Assistant Professor, Kathmandu University, Nepal
Mr. Pawan Karki	Assistant Professor, Kathmandu University, Nepal
Mr. Jahid Ahmed Jibran	Assistant Professor, Kathmandu University, Nepal
Mr. Uren Amatya	Lecturer, Kathmandu University, Nepal
Ms. Sirapa Shrestha	Teaching Assistant, Kathmandu University, Nepal
Mr. Sagar Pandit	Teaching Assistant, Kathmandu University, Nepal

Scientific Committee

Dr. Jeeban Poudel	University of Queensland, Australia
Dr. Milan Shrestha	Nanyang Technological University, Singapore
Dr. Anil Kumar Bastola	University of Nottingham, United Kingdom
Dr. Niroj Maharjan	Advance Remanufacturing and Technology Center, Singapore
Dr. Milan Paudel	Nanyang Technological University, Singapore
Dr. Arun Satyal	Nepal Engineering College, Nepal
Mr. Deepak Paudel	Aalto University, Finland
Mr. Janak Aryal	Wärtsilä Finland Oy
Dr. Bikash Adhikari	Kathmandu University, Nepal
Dr. Bishal Silwal	Kathmandu University, Nepal

Dr. Nawaraj Sanjel	Kathmandu University, Nepal
Dr. Sushil Shrestha	Kathmandu University, Nepal
Dr. Binay KC	Kathmandu University, Nepal
Dr. Krishna P. Shrestha	Kathmandu University, Nepal
Dr. Biraj Singh Thapa	Kathmandu University, Nepal
Mr. Biraj Dhungana	Green Road Waste Management, Nepal
Mr. Abhishek Karki	Abhiyantrika Karmashala, GDS Fiber World, Nepal
Dr. Parash Mani Timilsina	Kathmandu University, Nepal
Dr. Surendra Sujakhu	Kathmandu University, Nepal
Dr. Ranjit Shrestha	Kathmandu University, Nepal
Dr. Sailesh Chitrakar	Kathmandu University, Nepal
Mr. Pratisthit Lal Shrestha	Kathmandu University, Nepal
Mr. Malesh Shah	Kathmandu University, Nepal
Mr. Bijendra Shrestha	Kathmandu University, Nepal
Mr. Hari Dhakal	Kathmandu University, Nepal
Mr. Chiranjeevi Mahat	Kathmandu University, Nepal
Mr. Pawan Karki	Kathmandu University, Nepal
Mr. Gokarna Paudel	Kathmandu University, Nepal
Mr. Jahir Ahmed Jibran	Kathmandu University, Nepal
Mr. Nischal Chaulagain	Kathmandu University, Nepal
Mr. Uren Amatya	Kathmandu University, Nepal
Ms. Sirapa Shrestha	Kathmandu University, Nepal
Mr. Sagar Pandit	Kathmandu University, Nepal

Conference Session Chairs

Dr. Milan Poudel	Nanyang Technological University, Singapore
Dr. Anil Kumar Bastola	University of Nottingham, United Kingdom
Dr. Prajib Joshi	Western Regional Campus, Nepal
Dr. Bijay Thapa	EnergizeNepal, Nepal
Dr. Ranjit Shrestha	Kathmandu University, Nepal
Dr. Binaya K.C.	Kathmandu University, Nepal
Dr. Bishal Silwal	Kathmandu University, Nepal
Dr. Sailesh Chitrakar	Kathmandu University, Nepal
Mr. Biraj Dhungana	Green Road Waste Management, Nepal
Dr. Krishna Prasad Shrestha	Kathmandu University, Nepal
Dr. Shailendra Kr. Jha	Kathmandu University, Nepal
Dr. Nawaraj Sanjel	EnergizeNepal, Nepal

IOP Conf. Series: Materials Science and Engineering 1279 (2023) 011001 doi:10.1088/1757-899X/1279/1/011001

Conference Publication Editors

Dr. Surendra Sujakhu	Editor-in-Chief	Kathmandu University, Nepal
Mr. Prayush Jung Karki	Editor	Kathmandu University, Nepal
Ms. Anuapama Gaihre	Editor	Kathmandu University, Nepal

Supporting Partners

Department of Mechanical Engineering, Kathmandu University, Nepal

University Grant Commission, Nepal

Nepal Technology Innovation Center, Kathmandu University, Nepal

Green Hydrogen Lab, Kathmandu University, Nepal

Institute of Engineering, Tribhuwan University, Nepal

Nepal Astronomical Society, Nepal

Women of Aeronautics and Astronautics, Nepal

Nepal Society of Agricultural Engineers

Girls in Tech

Tekka Nepal

Society of Mechanical Engineering Students

Automobile and Mechanical Engineering Students' Innovative Nexus

Society of Mechanical Engineers' Nepal

Society of Innovative Mechanical Engineering Students

Co-hosted by: Himalayan University Consortium, ICIMOD

Hosted by: Kathmandu University

Organized by: Association of Mechanical Engineering Students

Peer Review Statement

All papers published in this volume have been reviewed through processes administered by the Editors. Reviews were conducted by expert referees to the professional and scientific standards expected of a proceedings journal published by IOP Publishing.

- **Type of peer review:** Single Anonymous
- **Conference submission management system:** Internal system used for KUSET (university online journal)
- **Number of submissions received:** 48
- **Number of submissions sent for review:** 24
- **Number of submissions accepted:** 10
- **Acceptance Rate (Submissions Accepted / Submissions Received × 100):** 20.8
- **Average number of reviews per paper:** 2
- **Total number of reviewers involved:** 1
- **Contact person for queries:**
  Name: Surendra Sujakhu
  Email: surendra.sujakhu@ku.edu.np
  Affiliation: Assistant Professor, Department of Mechanical Engineering, Kathmandu University

Comparative analysis of radiant and radiator heating system for a residential building

Ananta Aacharya[1], Robin Koirala[1*], Shuvas Khanal[1] and Bivek Baral[1]
[1] Department of Mechanical Engineering, Kathmandu University, Dhulikhel Kavre, Nepal
[*]Corresponding author: koiralarobin22@gmail.com

Abstract. Energy use for heating and cooling at the household level is rising annually. The residential sector consumes about 69% of energy consumption, out of which the highest percentage is shared by traditional fuel for cooking applications, followed by heating and cooling applications. In such a scenario, the primary goal should be to meet future heating and cooling energy demand through the use of novel technologies such as radiant heating systems (RHS). By utilizing low-temperature fluid flowing through the pipe, the RHS is more efficient than a standard heating system in maintaining the correct interior temperature. These systems are not frequently implemented in Nepal. It is crucial to compare the applicability of the system with technologies like radiator systems. The building provided by Urban Development and Building Construction (UDMC) is used for the study. The radiant and radiator heating system is designed, and the effects of different parameters on the heat transfer coefficient are studied. The maximum heating load is 50kWh for mid of January. A radiant heating system with a panel area of 18.24m^2 is sufficient to satisfy the heating demand, while radiator heating requires a panel area of 3.6m^2. To meet the same load, the total pipe length for the radiant and radiator heating systems is 300m and 42m, respectively. Radiant and radiator heating systems had a heat transfer coefficient of 7.76W/m^2 and 15.34W/m^2, respectively. The variation is because the radiator system must provide the same amount of heat to the building while having a smaller surface area than the radiant heating system.

Keywords: Heating Load, Radiant, Radiator, Performance, Heat transfer

1. Introduction

The world population is increasing, and people are coming out of poverty, demanding more housing options for living, schooling, and working. Building sectors utilize vast amounts of energy on a global scale to power their homes and workplaces, travel the world and operate factories that generate billions of dollars' worth of goods. Globally, the building industry consumes around 40% of total energy and emits more than 30% of CO_2 into the atmosphere [1]. Local pollution levels in cities have risen because of economic expansion and urbanization. This situation necessitates moving away from high-grade energy sources toward renewable energy sources, such as radiant heating systems for heating and cooling purposes [2]. Space heating via the radiant surface is a proven technology in Nordic countries and is used extensively in residential buildings. Almost all buildings in Korea have radiant floor heating installed (up to 95%); the installation number is 80% in Northern China [3].

Building energy consumption shares the most significant portion of the total energy demand in Nepal. Among these, a substantial amount of energy is used for cooking and only a tiny amount for heating and cooling application. With rising living standards and economic progress, people increasingly rely on electrical equipment for thermal comfort, such as air conditioners, heaters, and radiators. Around 10%

of home energy use in Nepal is used for heating and cooling applications, and the consumption rate is increasing annually [4]. Thus, to meet the future heating and cooling energy demand in the building sector, radiant heating might be an appropriate and urgent technology.

Solar-assisted floor heating (SAFH) comprises a solar collector, thermal storage, water pump, control valve, control system, and radiant surface, also known as the active layer [5]. A radiant surface is a heat-emitting surface that exhibits complicated behaviors due to the coexistence of all modes of heat transport and thermal capacitance [6]. Additionally, the system might be combined with an auxiliary heating system for a reliable energy supply. Sometimes, the SAFHS is combined with a ground source heat pump with a passive heating system to get optimal benefits from the system. Solar integrated combi-system has a minimal operating cost [7], higher electric COP [8], higher energy and exergy efficiency [9], and better thermal performance [10].

Similar to the radiant heating system (RHS), radiator heating is often utilized for space heating. Unlike radiant heating, the heat transfer in the radiator system is convection dominant; the ratio of radiation to convection is under 50%, and the value is 10% for a radiator with metal fins [11]. The radiator heating systems provide several advantages, including easy maintenance, low/minimum running costs, dependability, and conductivity. The convective nature assists in lowering the interior air humidity, but the heating mechanism is sensible. The heat transfer area of the radiator, fluid temperature, and flow rate affect the heat flux produced by the system. The placement of radiators in a heating system has a significant impact on thermal comfort and temperature dispersion [12]. The space should have an adequate ventilation system installed since the surface temperature is sufficiently high, igniting dust particles and resulting in poor indoor air quality. Other downsides of radiators include noise, the need for adequate ventilation, and a high surface temperature.

The purpose of this study is to evaluate two systems, radiant and radiator, in terms of their size, performance, efficiency, and material composition. The first section contains a calculation of the heating load using TRNSYS as well as the determination of the optimal system size, whereas the second portion analyzes the heat transfer coefficient and the factors impacting the phenomena by modelling the system using fundamental heat transfer equations.

2. Methodology

2.1. Building Description

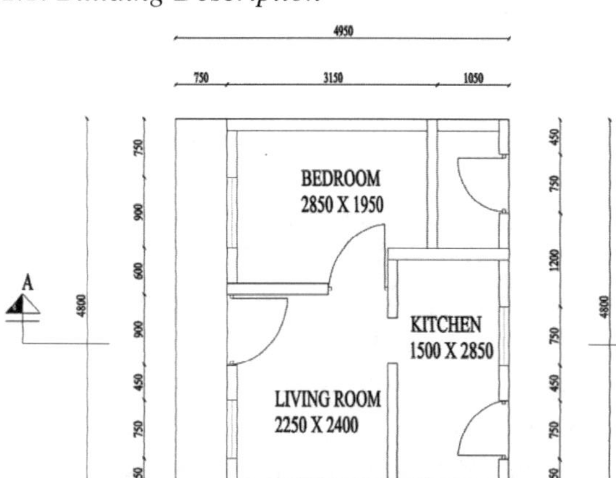

Figure 1. Floor plan of the building

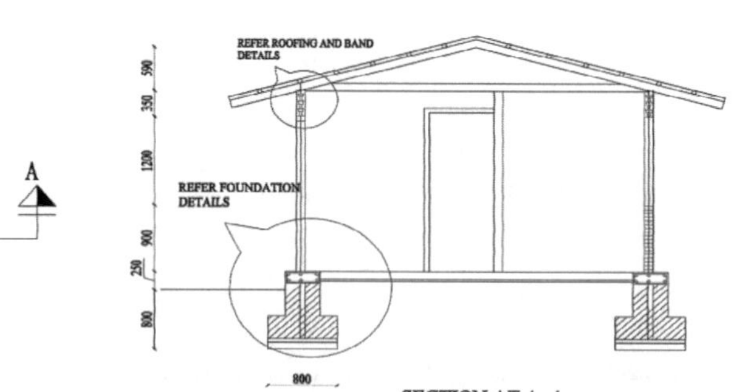

Figure 2. Section view of the building

The thermal system is designed and simulated for 23.76m^2 residential building provided by the UDMC. The building is constructed of stone masonry and has a structural attic roof. Stone is used to build both the internal and external walls, with cement mortar used to complete the inside. The base is made of

RCC, and the upper surface is covered in 20mm thick wood flooring. This home has a kitchen, living room, bedroom, and bathroom and is made to fit a nuclear family of four. The bedrooms and living rooms are larger than the others to accommodate guests and relatives, while the smaller kitchen is in the northeast corner. The floor height is 2.45m, and the detail of the floor plan is depicted in Figures 1 and 2. For simplicity, the living room, bedroom, kitchen, and bathroom are labelled as Zone 1, Zone 2, Zone 3, and Zone 4, respectively.

2.2. Building Materials

All building supplies are selected based on their popularity and accessibility in Nepalese society. The building had a brick masonry wall and is one story tall. The study considered a single piece of transparent glass that is 6mm thick, 1.2m tall, and 0.75m long. The information on the building components and corresponding overall heat transfer coefficient used for simulation and further analysis are summarized in Table 1.

Table 1. U-value for components of building

Building Component	Materials	Thickness	U-value[W/m^{2}K]
External Wall	• Plaster board • Brick masonry	220mm	1.308
Internal Wall	• Plaster board • Brick masonry • Plaster Board	240mm	1.124
Ceiling	• Plywood	20mm	3.297
Floor	• Timber wood • Insulation	20mm	0.039
Roof	• GI sheet		4.346
Window	• Single glazing	25mm	5.68
Door	• Plywood	20mm	3.297

2.3. Model in TRNSYS

The heating load of the building is computed using the TRNSYS 16 program, a transient system simulation software intended to estimate buildings' energy performance by including meteorological data, as shown in Figure 3.

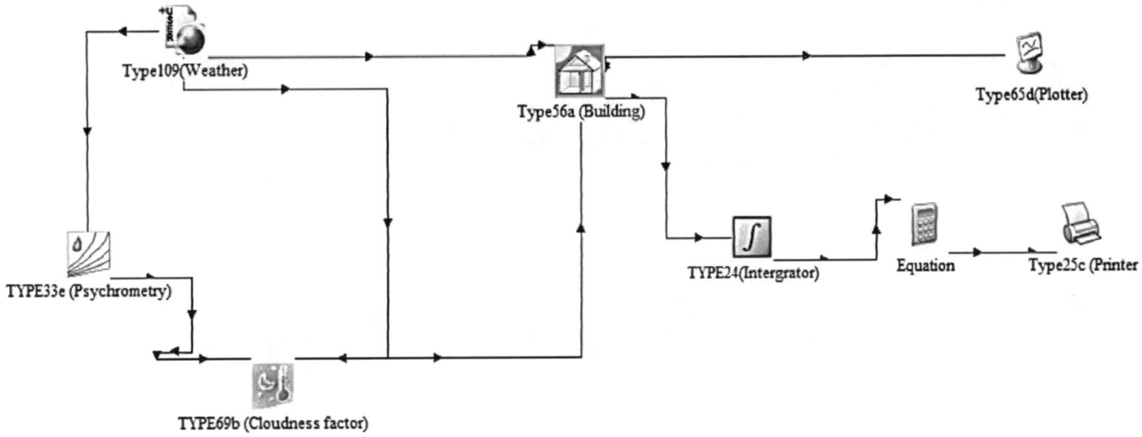

Figure 3. The links before simulation in TRNSYS studio

IOP Conf. Series: Materials Science and Engineering 1279 (2023) 012001 doi:10.1088/1757-899X/1279/1/012001

The building is modelled using TRNSYS type56a, which can create structures such as walls, windows, roofs, and floors, as well as the proportional thickness of each layer inside the structure. Five distinct zones are established, each with its own area and volume defined by the architectural design. Type 33e receives the moist air's dry bulb temperature and relative humidity as input and outputs the moist air's characteristics and enthalpy. Type 69b is used to measure the effective sky temperature. Type 69b is also used to compute the long-wave radiation exchange between an arbitrary external surface and the atmosphere. The weather and meteorological data are provided for the case of Dhulikhel, Nepal by Type 109. It aids in determining the ambient temperature, wind velocity, and intensity of solar irradiation. Selected system variables are generated at predefined time intervals using type 25c. The plotter simulator is Type 65d, which displays selected system variables while the simulation progresses. The collective information related to weather data, infiltration, equipment loads, and thermal properties of building components are shown in Table 2.

Table 2: Parameters used in the simulation with their description.

S.N	Parameters	Description
1	Weather data	Meteonorm weather file of Kathmandu is taken into consideration.
2	Infiltration	0.5 ACH is considered as suggested by ASHRAE.
3	Equipment load	Equipment heating load is not taken into consideration.
4	Windows	Single-clear glass of U-value (5.68 W/m °C)
5	Heating Schedule	Bedroom (20:00 to 6:00) Living room (20:00 to 6:00) Kitchen (6:00 to 9:00) & (18:000 to 21:00) Toilet (6:00 to 9:00) & (18:00 to 20:00)

2.4. System Description

2.4.1. Radiant Heating system

The solar-assisted-radiant floor heating system consists of several components: collector system, thermal storage system, pumping system, and radiant floor heating system, as shown in Figure 4. The absorbent material absorbs solar energy in flat plate collectors, which raises the temperature of the supply water from T_1 to T_2. The thermal storage tank stores the heated water at a temperature T_s. The same hot water flows through the heating coil and increases the room temperature by convection and radiation heat transfer. The collector has an adequate size that can meet all the household hot water demands on sunny days. On cloudy, rainy, or snowy days, the thermal tank's electric heating rods are activated, providing the required hot water for the heating system.

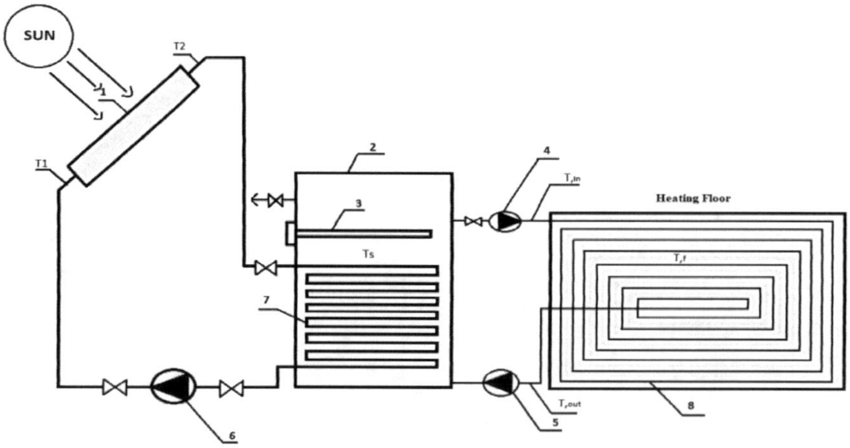

1- Flat plate collector 2-Thermal storage tank 3-Heating rod 4,5(6)-Pump 7-Heat exchanging coil 8-Heating coil

Figure 4. Flow diagram of solar-assisted floor heating system

IOP Conf. Series: Materials Science and Engineering 1279 (2023) 012001 doi:10.1088/1757-899X/1279/1/012001

2.4.2. Radiator Heating system

In this system, hot water produced by solar thermal collectors is initially kept in thermal storage during the hours of the sun. The hot water is subsequently circulated via the radiator (heat exchanger unit), which distributes the heat throughout the room. As shown in Figure 5, the water pump constantly circulates the hot water (heat carrier) throughout the heating circuit. When the stored temperature is less than the setpoint temperature, the auxiliary heater attached ahead of thermal storage warms water during off-hours.

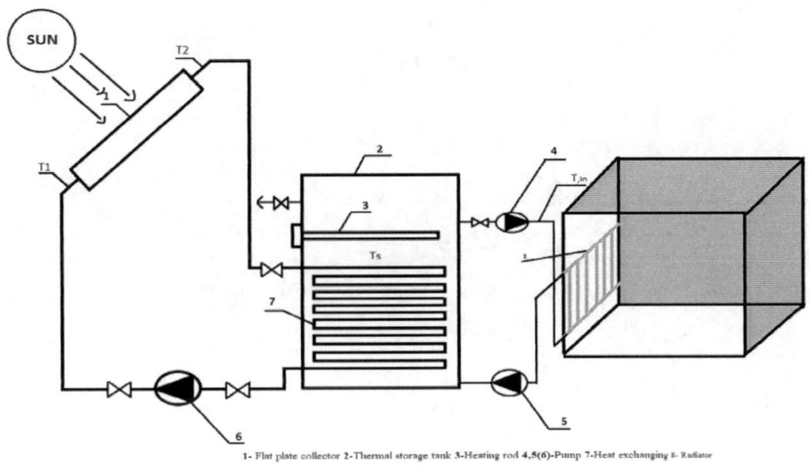

Figure 5. Schematic Layout of Radiator Heating System

2.5. System Sizing

2.5.1. Solar Collector

The collector type and the expected temperature difference significantly affect the system's overall performance. The collector and thermal storage design technique for the radiant heating and the radiator heating system is the same; the only difference is the design parameters' value. The collector solar heat production is calculated using solar insolation over a surface tilted at β (kWh/m^2 day) (I), the total collector area (A_c) and collector efficiency (η_c) [13] by:

$$Q_s = A_c \times I \tag{1}$$

The useful heat production of a flat plate collector can be calculated by energy balance in the fluid volume and is given as:

$$Q_u = m_c \times c_p \times (T_2 - T_1) \tag{2}$$

Where m_c is the fluid flow rate, c_p is the specific heat, T_1 and T_2 are the inlet and outlet temperature. The thermal efficiency of the collector is evaluated using useful heat production and solar heat production as:

$$\eta_{th} = \frac{Q_u}{Q_s} \tag{3}$$

The energy required for the floor heating system is

$$E = Q_L \times t \tag{4}$$

Where, Q_l is the heating load of the building t is running time.

The solar collector must supply the energy required for floor heating. The mass of water flowing through the collector is calculated [14] by:

$$Q_u = m_c \times c_p \times (T_2 - T_1) \tag{5}$$

2.5.2. Thermal Storage

The storage capacity depends on the storage temperature, the thermophysical properties of fluids like specific heat capacity, and the density of the material. This study considered water as a working fluid because of the high specific heat capacity and availability. The volume of the storage tank is calculated using the value of following input parameters [15]:

- total heat collected by the collector (Q_c)
- the specific heat capacity of the water in the tank (C_p)
- the density of the water in the tank (ρ),
- the utilization of temperature difference between the heat supply temperature and the average temperature at the inlet to floor heating system (ΔT).

$$V = \frac{Q_c}{C_p \times \rho \times \Delta T} \tag{6}$$

The heat collected by the flat plate collector is given by

$$Q_c = R_t \times \eta_{th} \times A_c \tag{7}$$

Where R_t is the total radiation on the plane of the solar collector KJ/m^2.

2.6. Heat transfer analysis

2.6.1. Radiant Heating

The Dittus and Boelter equation is used for internal flow, and natural convection turbulent flow is considered for the radiant surface. In the case of floor heating, the inlet and outlet temperature of hot fluid are known variables; also, the room temperature is fixed at a specific temperature. This problem can be solved by Log Mean Temperature Difference (LMTD). A simple relation of the given form can calculate heat transfer.

$$Q = U \times A \times \Delta T_{lm} \tag{8}$$

Where, U= Overall heat transfer coefficient
A= area of heat transfer
ΔT_{lm}=Log Mean Temperature Difference

The overall heat transfer coefficient is given by:

$$U = \frac{1}{A \sum R_{Th}} \tag{9}$$

Where, A= Heat transfer area
R_{Th}= Total thermal resistance given as:

$$R_{th} = \frac{1}{\frac{1}{h_i} + \frac{L}{K} + \frac{1}{h_o}} \tag{10}$$

Where, h_i, h_o inside and outside heat transfer coefficient
L = the thickness of the slab
K= conductivity of concrete slab

For this problem inside heat transfer coefficient is calculated by using Dittus and Boelter equation

$$Nu = 0.023 \times {Re_D}^{0.8} \times Pr^{0.3} \tag{11}$$

An outside heat transfer coefficient is calculated by using an external flow problem with constant surface temperature and turbulent flow.

$$Nu = 0.10 \times (GrPr)^{0.333} \tag{12}$$

Where Gr= Grashof number is given by:

$$Gr = \frac{g \times \beta \times \Delta T \times {L_c}^3 \times \rho^2}{\mu^2} \tag{13}$$

Where, symbols have their usual meaning
Thermal resistance due to pipe thickness is neglected as thickness is considerably small and conductivity is high.

The LMTD for the counter-flow heat exchanger is given by:

$$\Delta T_{lm} = \frac{(T_{h2} - T_{c1}) - (T_{h1} - T_{c2})}{\ln(\frac{T_{h2} - T_{c1}}{T_{h1} - T_{c2}})} \tag{14}$$

Where, T_{h1}, T_{h2} the hot fluid inlet and outlet temperature
T_{c1}, T_{c2} the cold fluid inlet and outlet temperature

2.6.2. Radiator Heating
The heat transfer from the radiator to the room is by natural convection. All sets of equations used for the airside are convection equations, and that for the water side is Dittus and Boelter equation. Newton's Law of cooling gives heat transfer from the radiator to the room.

$$Q = h \times A \times \Delta T \tag{15}$$

Where ΔT = temperature difference between system and surrounding
A = Surface area of heat transfer
h = heat transfer coefficient
Mathematically, heat transfer coefficient h is calculated from the Nusselt number, which is again the function of Rayleigh's number is given by:

$$Nu = \frac{h \times L}{k} \tag{16}$$

Where, L = characteristic length of radiator
k = conductivity of fluid at the film temperature
h = heat transfer coefficient

2.6.3. NTU method
The radiator outlet temperature is unknown, and the heat transfer area of the radiator system is the subject of study. For the proper design of the system, the length of the pipe must be known. Thus, this problem is solved by the NTU method. The effectiveness of the system is given by:

$$\in = \frac{C_c(T_{co} - T_{ci})}{C_{min}(T_{hi} - T_{ci})} \tag{17}$$

The heat transfer from the radiator is given by:

$$Q_{actual} = \in \times q_{max} \tag{18}$$

Since the fluids are unmixed and flowing of air is perpendicular to the radiator. For this case, the NTU is given by:

$$N = \frac{-1}{C} \times [1 + C \times \ln(1 - \in)] \tag{19}$$

Where, $C = \frac{C_{min}}{C_{max}}$

The area of the heat exchanger is calculated as follows:

$$A = \frac{N \times C_{min}}{U} \tag{20}$$

Where U = overall heat transfer coefficient
For this problem inside heat transfer coefficient is calculated by using Dittus and Boelter equation:

$$Nu = 0.023 \times {Re_D}^{0.8} \times Pr^{0.3} \tag{21}$$

3. Result and Discussion

3.1. Heating Load Calculation

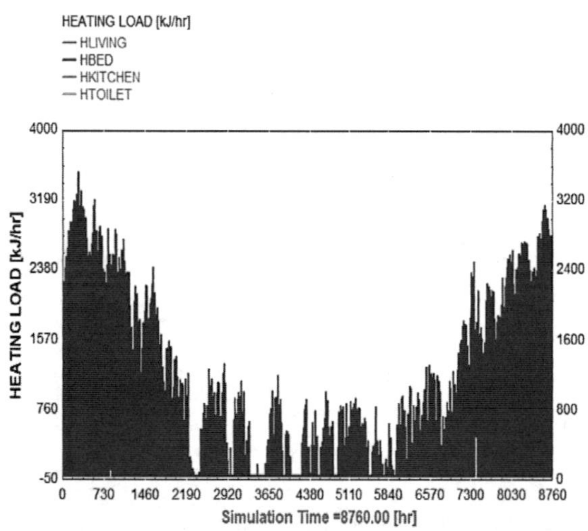

Figure 6. Simulated heating load for bedroom

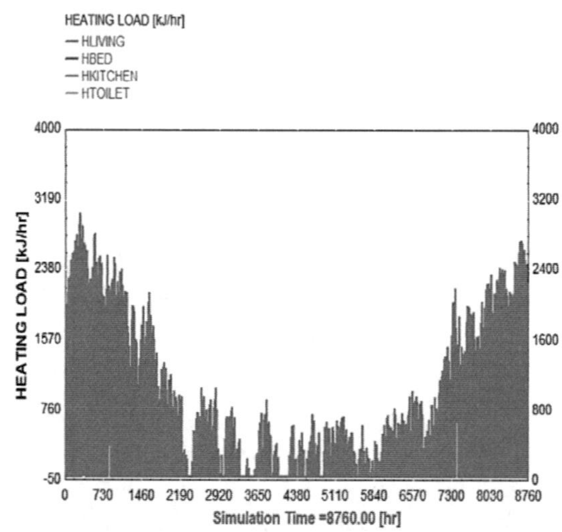

Figure 7. Simulated heating load for Living room

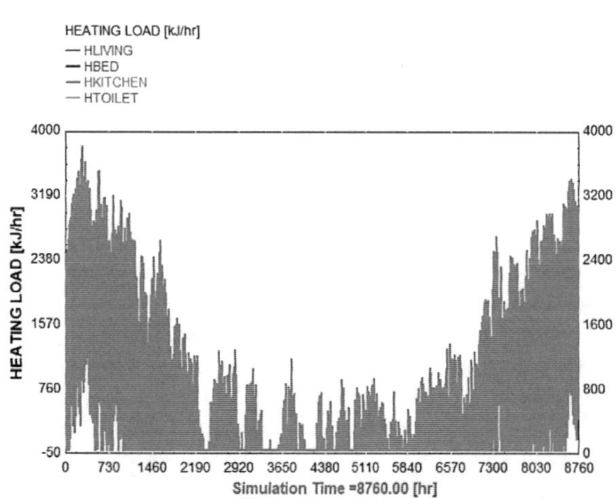

Figure 8. Simulated heating load for kitchen.

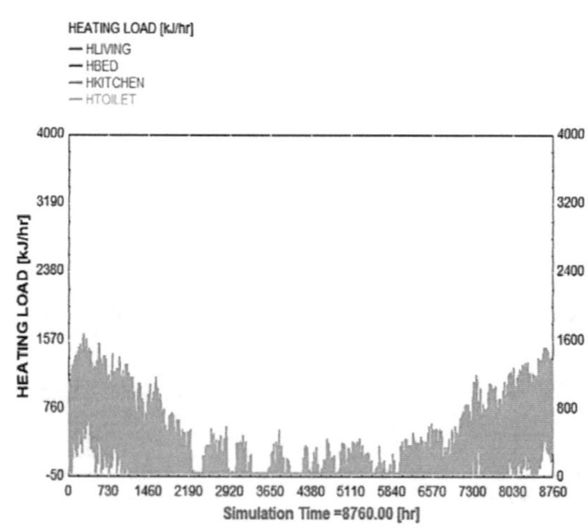

Figure 9. Simulated heating load for toilet

Figures (6-9) present the simulation results on the monthly heating requirement by zone for a building in Kathmandu. The results show that the heating demand is maximum for the kitchen and bedroom. The size, characteristics, and orientation of the building components affect the heating load. The bedroom is the largest room, while the kitchen is situated to the northeast. Figure 6 illustrates the hourly change in bedroom load, which has a peak demand of around 3500kJ/hr in January. The same demand is about 4000kJ/hr for the kitchen, as depicted in Figure 8. The statistics showed that January has the highest

IOP Conf. Series: Materials Science and Engineering 1279 (2023) 012001 doi:10.1088/1757-899X/1279/1/012001

demand in each zone and that the year's beginning and last four months have the highest heating requirements. For four months, starting from April to July has minimum heating requirements of about 500KJ/hr.

Figure 10 shows the monthly energy demand for the building located in Kathmandu. Consequently, the heating demand for May is less than 50kWh, and the building' s heating demand is much reduced for three months. Heating and infiltration loads are at their peak in January with values of 1039kWh and 90.31kWh, respectively. It is common to size heating and cooling systems for the worst case so that the system can meet the demand for the worst condition accordingly. The heating demand of the building is maximum for January, as depicted in the figure. The building has maximum daily heating demand of 51.51kWh in the middle of the month, as shown in Figure 11.

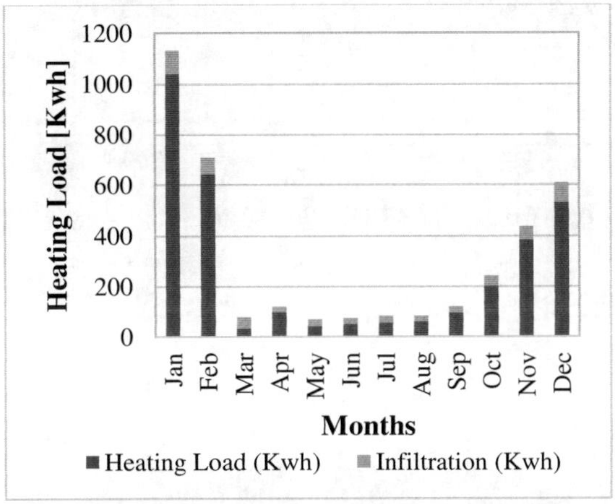

Figure 10. Cumulative monthly Heating load of the building in Kathmandu

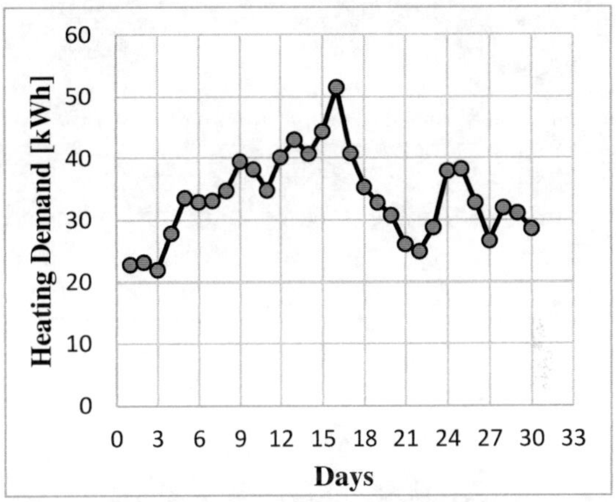

Figure 11. Daily Heating demand in (kWh)

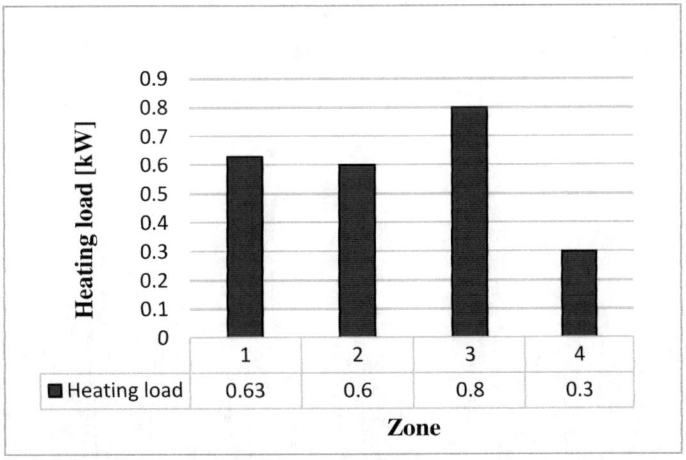

	1	2	3	4
Heating load	0.63	0.6	0.8	0.3

Figure 12. Maximum Hourly Zone wise heating demand

3.2. Zone Temperature

Figure (13-16) shows the room temperature variation throughout the year. The TRNSYS simulation is conducted with a time step of one hour. The attic temperature of the building is slightly higher than the ambient temperature of Kathmandu valley. This variance occurs because of the heated air stored in the attic zone, which raises the total attic temperature. Figures (13 and 14) depict the zone temperatures; the minimum temperature is around 2°C. The room temperature should be kept at 20°C; hence, the building requires a heating system capable of maintaining a 20°C zone temperature. The hourly temperature variance reveals that the kitchen and toilet are consistently cold throughout the year. As mentioned previously, the building's orientation influences zone temperature, affecting the heating and cooling load. The kitchen's exterior wall faces north and east, whereas the toilet room faces north and west. These are the sole factors that contribute to temperature fluctuation within these zones.

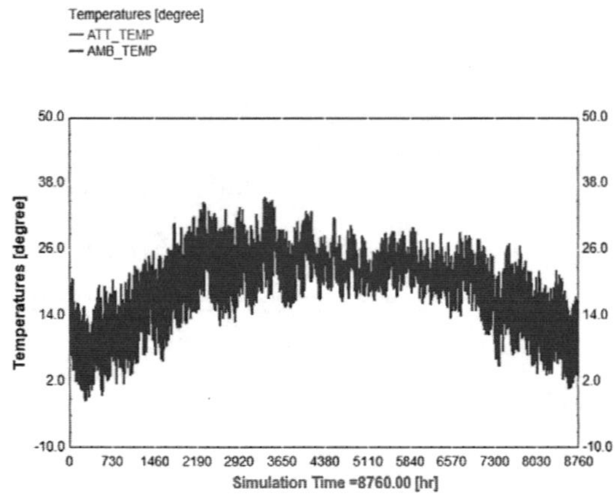

Figure 13. Ambient temperature for Kathmandu

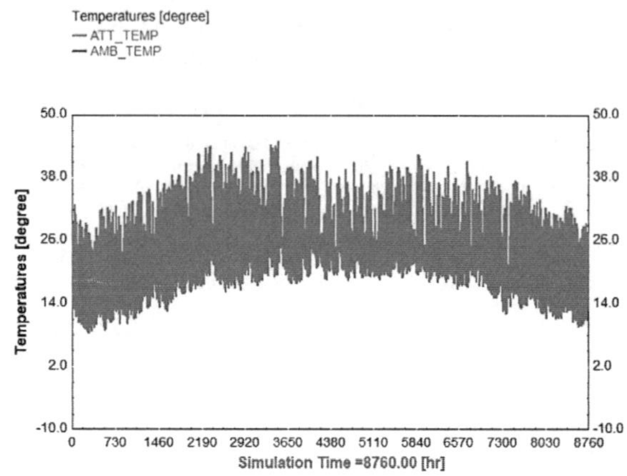

Figure 14. The attic temperature for the building

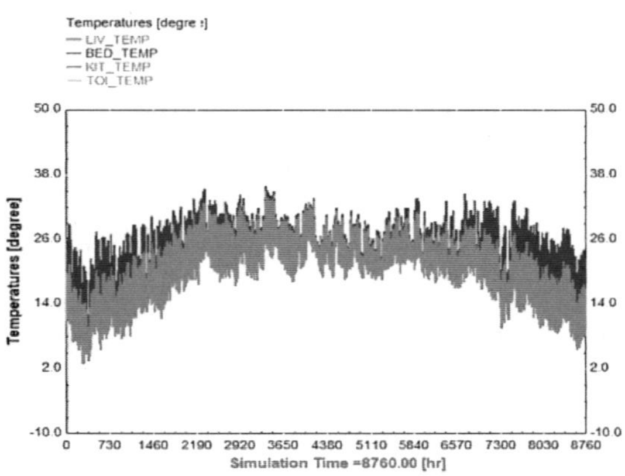

Figure 15. Bedroom and Kitchen temperature

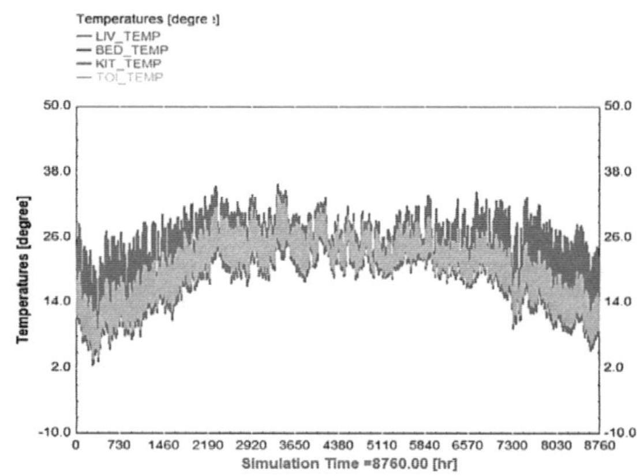

Figure 16. Living room and toilet temperature

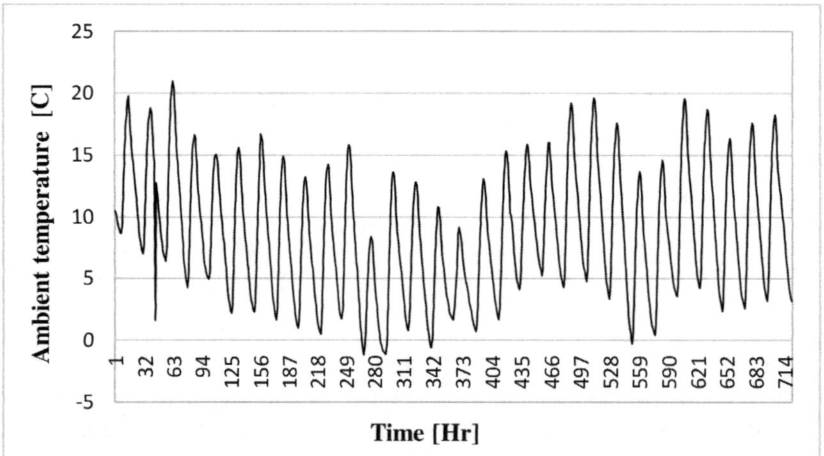

Figure 17. Hourly temperature variation for January

The result indicates that for about four months, the minimum temperature is greater than the room's setpoint temperature. Between May and August, the building is entirely self-sufficient in terms of heating. However, the average low temperature is lower than the setpoint temperature at the start and end of the year, necessitating the use of a heating system during these months. Ambient temperature is essential for heat transfer analysis. Figure 17 depicts the hourly ambient temperature for January. The lowest temperature possible is -0.79°C, which does not occur frequently. As a result, the average minimum ambient temperature is set at 30°C.

3.3. Sizing the System

The daily heating requirement for the building is 52kWh. For Kathmandu, the average solar insolation is 4.7 kWh/m^{2}day [16], and the solar radiation intensity is 777.27kW/m^2 [17]. It is mentioned that for the highly efficient operation of solar floor heating systems, the water temperature range should be 40°C to 50°C [18]. The collector outlet temperature is estimated to be at least 60°C. The efficiency of the collector is assumed to be 0.50. A collector area of 18.5m^2 is necessary to satisfy the daily heating requirement. The heat collected by the flat plate collector is estimated first, and then the storage tank's capacity is determined, which is found to be 5.15m^3 (Table 3). The radiant heating system requires a total of 300m of 13mm diameter tubing. Table 4 includes other factors such as source temperature, surface temperature, and flow rate.

Table 3. Value of parameters required for system sizing

S.No.	Parameters	Symbol	Results	
			Radiant Heating	Radiator Heating
1	Ambient high temperature	T_a	10	10
2	Ambient water temperature	T_1	19	19
3	Fluid outlet temperature	T_2	60	85
4	Stored Water temperature	T_s	55	80
5	Heating Load	Q_l	52	52
6	Collector area	A_c	18.5	25
7	Storage volume capacity	V_s	5.51	6.5

Table 4. Designed parameters for radiant heating system

S.No.	Parameters	Zone Wise Value				Mean Value
		Zone 1	Zone 2	Zone 3	Zone 4	
1	Tube Size (mm)	13	13	13	13	13
3	Supply Temperature	54.9	48.2	55	54.9	53.25
4	Floor Temperature	32	29.7	32	32	31.42
5	Flowrate (lit/sec)	0.56	0.37	0.66	0.18	
6	Loop length (m)	88.16	91.2	88.16	30.4	300[a]

3.4. Parametric Study

Based on simulation results, it is determined that a 3KW heating system could fulfill the building's heating needs. The authors evaluated the influence of the mean fluid temperature and velocity of the fluid on the system's total heat transfer coefficient. As depicted in Figure (18-21), the overall heat transfer coefficient changes linearly with varying mean fluid temperature and significantly increases with velocity.

The impact of mass flow rate on temperature difference is studied. The result (Figure 20) shows that increasing the mass flow rate reduces the temperature difference between fluid entering and leaving the heating circuit. When the mass flow rate is increased, the water exits at a greater temperature, potentially carrying thermal energy. As a result, circuit efficiency is reduced since the potential thermal energy carried by the fluid exiting the circuit is unused. Decreasing the mass flow rate reduces heating pipe

IOP Conf. Series: Materials Science and Engineering 1279 (2023) 012001 doi:10.1088/1757-899X/1279/1/012001

diameter, allowing smaller pipe diameters with reduced pipe spacing to compensate for the decrease in room temperature.

The effect of the mass flow rate of air on actual heat transfer is investigated. As depicted in Figure 21, the actual heat transfer varies linearly with varying air mass flow rates. However, the airflow rate is constrained by thermal comfort, as air velocity within the conditioned space has an effect on the air temperature distribution. Because the heat transmission mechanism is natural convection, the suggested flow rate should be reduced.

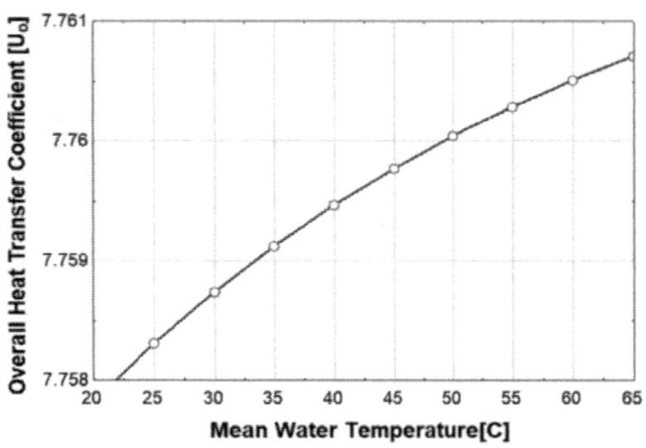

Figure 18. Graph of the mean temperature of fluid vs. heat transfer coefficient

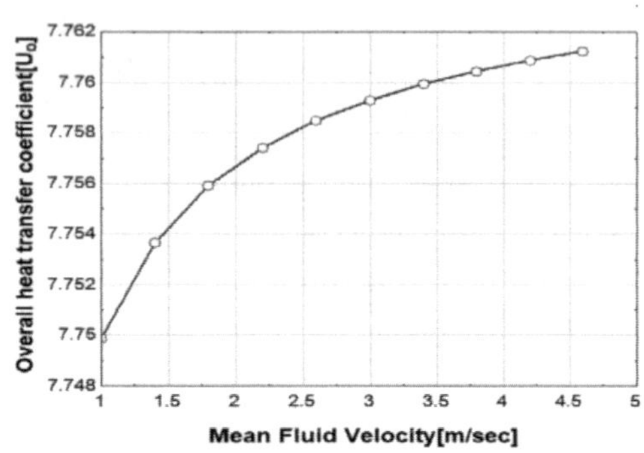

Figure 19. Graph of mean fluid velocity vs. overall heat transfer coefficient

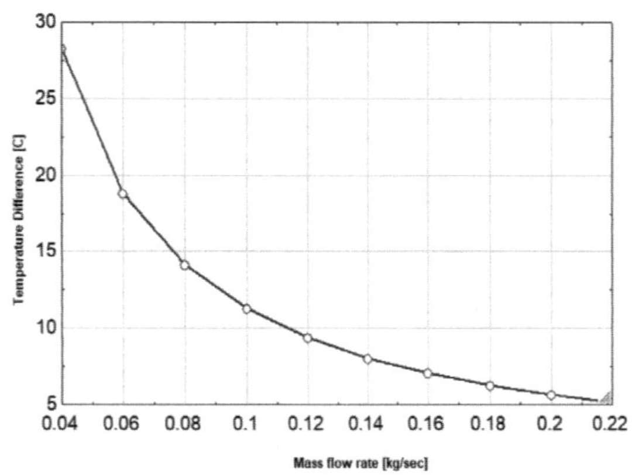

Figure 20. Effect of mass flow rate on the temperature difference of fluid

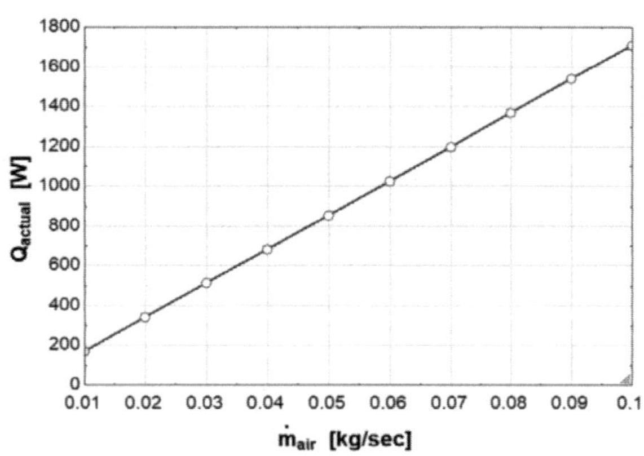

Figure 21. Graph of actual heat transfer versus mass flow rate of air

3.5. Comparative Analysis

A comparative study is performed based on system size and heat transfer analysis. Radiant heating is a low-temperature technology that provides thermal comfort, improved performance, and long-term cost savings. Solar-assisted radiant/radiator systems utilize the same heating units and components as conventional radiant/radiator systems. They are identical except for the heat-emitting device. Radiant heating systems use a radiant surface to generate heat and a heat exchanger in the radiator. Heat exchangers come in various shapes and sizes and may include a finned surface to aid in heat transfer. As shown in Table 5, a radiant heating system requires approximately 18.5m^2 collector area and 5.15m^3 storage volume. For the same design condition radiator system requires 25m^2 collector and 6.5m^3 storage tank. This is because radiator heating requires a high temperature working fluid to provide the same level of warmth. At peak solar radiation, the highest collector efficiency is around 0.5 for radiant heating systems and 0.45 for radiator heating systems, with respective collector gains of 105894kJ and 130050kJ. The heat transfer study is conducted using the parameters of the building's integrated heating system.

Water is used as the working fluid and has an inlet temperature of 55°C; it flows at a rate of 0.37kg/sec through a PEX pipe of 13mm outer diameter. Using Dittus and Boelter's equation for internal flow and natural convection with the turbulent flow for external flow gives an overall heat transfer coefficient of about 8W/m2K. The heat exchanger has a diameter of 25mm without a fined surface and a flow rate of 0.1kg/sec. It is filled with fluid with an intake temperature of 80°C. Under calculation, a total of 3.6m2 heating area with a pipe length of 40m can emit heat of about 4000W. The overall heat transfer coefficient for the same system is 15.34W/m2K. This value is higher than that of radiant heating system, but it is still considered to be low because the fin element was not considered in this study. However, the system's 4000W output is sufficient to heat the building to the required level.

Table 5. Comparative table for the radiant and radiator heating system

S.No.	Parameter	Results	
		Radiant Heating	**Radiator Heating**
1	Collector area	18.5 [m^2]	25 [m^2]
2	Collector heat gain (peak solar radiation)	105894 [KJ]	130050 [KJ]
3	Storage Volume	5.15 [m^3]	6.5 [m^3]
4	Collector efficiency (maximum value)	0.51	0.45
5	Diameter of pipe	0.013 [m]	0.025 [m]
6	Inlet temperature	55 [°C]	80 [°C]
7	Mass flow rate	0.66 [kg/sec]	0.1 [kg/sec]
8	Heat transfer area	18.24 [m^2]	3.6 [m^2]
9	Length of pipe	300 [m]	42 [m]
10	Overall heat transfer coefficient	7.76 [W/m^{2}K]	15.34 [W/m^{2}K]
11	Actual heat transfer	3000 [W]	4000[W]

4. Conclusion

Due to economic growth and improved living standards, there is a sense of urgency for heating and cooling systems for thermal comfort. Although the market offers a variety of heating and cooling solutions, each of them has its pros and cons. Considering the thermal comfort, energy consumption, and environmental effects, researchers and engineers must concentrate on developing novel heating and cooling systems. Radiant heating systems, which are particularly popular in the Nordic nations, have the potential to be an excellent future heating and cooling option for Nepalese structures. The purpose of this work is to compare the hydronic radiant floor heating system with the radiator heating system for a residential building with a floor area of 23.76m^2. The heating load of the building is calculated using a transient system simulation program, TRNSYS 16, which is used to simulate the energy performance of the building by introducing the meteorological data of Kathmandu, Nepal.

The key findings of the study are stated as follows:

- The building provided by Urban Development and Building Construction is used for the study, where the maximum heating load for the building is 50kWh for mid-January.
- The radiant and radiator heating systems are designed, and different parameters are studied to know the impact on the overall heat transfer coefficient. The collector of 18.5m^2 for radiant and 25m^2 for radiator can meet the heating demand of the building. The overall heat transfer coefficient for the radiant panel and radiator is 7.76W/m^2 and 15.34W/m$^{2.}$ respectively. To sum up, the radiator heating system has a higher heat transfer coefficient because the unit must deliver the same load for the building with a smaller surface area than the radiant heating system.
- The radiant heating system has a panel area of 18.24m^2 with a spiral-shaped pipe length of 300m embedded under the floor surface. In contrast, the radiator system requires only a heat exchanger of about 3.6m^2 to meet the same heating demand of the building with a pipe length of 42m.

- The mass flow rate on temperature difference is studied, and the result shows that increasing the mass flow rate increases the temperature of fluid leaving the heating circuit. This causes a reduction in circuit efficiency because the potential power is not being utilized. Overall, it is concluded that Low-flow rate systems are energy-efficient systems.

5. Refrences

[1] Yang L, Yan H and Lam J C 2014 Thermal comfort and building energy consumption implications – A review *Applied Energy* **115** 164–73

[2] Connolly D, Lund H, Mathiesen B V, Werner S, Möller B, Persson U, Boermans T, Trier D, Østergaard P A and Nielsen S 2014 Heat Roadmap Europe: Combining district heating with heat savings to decarbonise the EU energy system *Energy Policy* **65** 475–89

[3] Bean R, Olesen B W and Kim K W 2010 Part 2 history of radiant heating & cooling system *ASHRAE Journal* **52** 50-55

[4] Shrestha S M 2017 *Energy efficient building for Nepalese market* thesis (Institute for Building, Energy & Material Technology, UiT The Arctic University of Norway)

[5] Mehdaoui F, Hazami M, Messaouda A and Guizani A 2020 Performance analysis of two types of Solar Heating Systems used in buildings under typical North-African climate (Tunisia) *Applied Thermal Engineering* **165** 114203

[6] Nielsen L 2014 *Analysis of the radiant heating and cooling System in the Green Energy Laboratory* thesis (Department of Energy and Process Engineering, Norwegian University of Science and Technology, Norway)

[7] Bellos E and Tzivanidis C 2017 Energetic and financial sustainability of solar assisted heat pump heating systems in Europe *Sustainable Cities and Society* **33** 70–84

[8] Zhai X Q, Yang J R and Wang R Z 2009 Design and performance of the solar-powered floor heating system in a green building *Renewable Energy* **34** 1700–8

[9] Atmaca I and Kocak S 2014 Theoretical energy and exergy analyses of solar assisted heat pump space heating system *Thermal Science* **18** 417–27

[10] Cho J, Park B and Lim T 2019 Experimental and numerical study on the application of low-temperature radiant floor heating system with capillary tube: Thermal performance analysis *Applied Thermal Engineering* **163** 114360

[11] Sarbu I and Sebarchievici C 2016 Performance Evaluation of Radiator and Radiant Floor Heating Systems for an Office Room Connected to a Ground-Coupled Heat Pump *Energies* **9** 228

[12] Plytaria M T, Tzivanidis C, Bellos E and Antonopoulos K A 2018 Energetic investigation of solar assisted heat pump underfloor heating systems with and without phase change materials *Energy Conversion and Management* **173** 626–39

[13] Mishra S K, Bajracharya T R and Bhattarai R N 2017 Design and Numerical Analysis of Solar Underfloor Heating System: A Case Study of Resort in Nagarkot, Nepal *Proceedings of IOE Graduate Conference* **5** 529–36

[14] Wu X, Liu Y, Liu G, Wang F and Wang Z 2017 Effect of Supply Air Temperature on Indoor Thermal Comfort in a Room with Radiant Heating and Mechanical Ventilation *Energy Procedia* **121** 206–13

[15] Aryal A and Bhattarai N 2017 Performance Analysis of Solar PV System of Teaching Hospital, Kathmandu, Nepal *Proceedings of IOE Graduate Conference* **5** 23–9

[16] Poudyal K, Kumar Bhattarai B, Sapkota B and Kjeldstad B 2012 Solar Radiation Potential at Four Sites of Nepal *Journal of the Institute of Engineering* **8** 189–97

IOP Conf. Series: Materials Science and Engineering 1279 (2023) 012001 doi:10.1088/1757-899X/1279/1/012001

[17] EEM Energy & Environment Media GmbH 2017 EEM Energy & Environment Media GmbH i. size is responsible for the SolarServer *Solarserver*

Numerical study of a coil pump and comparison with field measurement

A Pandey[1*], P L Bijukchhe[1], P Khanal[1], S Acharya[1], S Chitrakar[1] and H P Neopane[1]

[1]Department of Mechanical Engineering, Kathmandu University, Nepal
[*]Corresponding address: arun.pandey@ku.edu.np

Abstract. Rural irrigation could be challenging in regions without proper access to electricity. In many remote places of Nepal, where the irrigation takes place in the vicinity of a river stream, a river water pump or coil pump can be used to supply water from the lower river terrain to the field. Water powered coil pump is a reciprocating pump that operates by rotating a coiled tube, by harnessing the power of a flowing river. A comparison in performance of cylindrical and conical shaped coil pumps was done both experimentally as well as numerically in this work. Under the river velocity of 0.9m/s, the maximum discharge at zero lift condition for the conical pump was 11,500 liters/day while for the cylindrical pump it was 8,600 liters/day. However, the conical pump could not generate lift more than 6.5m while the cylindrical pump performed well even at 11m head. The Numerical study was also carried out in order to visualize the flow patterns and compare the performance of the pumps while varying different parameters. The 3D computation domain of the pump for four cases (two cases varying the number of turns and two cases with change in shape) were generated and numerically studied using *OpenFOAM*. The interaction, being a multiphase case, *interFoam* solver was used, and to simplify the simulation laminar turbulence model was taken. The pressure and water volume at the outlet in the simulation were measured and compared between different cases. It was observed that the pressure obtained while using a cylindrical pump was 2.52 % more than that while using the conical pump. The outlet pressure was found to be increasing by 35% while increasing the number of turns from 7 to 15 in cylindrical and 42.5% in case of conical coil pump. The volume flow rate was observed to be more frequent in the case of the conical pump. Due to constraints of computational conditions, the experimental setup could not be totally replicated. However, the outcome showed a similar trend as obtained in the experiment. It was found that the cylindrical pump was better for the higher head, whereas the conical pump had shown better results for higher discharge.

Keywords: Coil Pump, Rural Irrigation, CFD, OpenFOAM

1. Introduction

The problems of irrigation in the hilly region of Nepal are very common ever since the start of human settlement in the region. This is due to the height difference between farming land and water source. The available water source, like the river, flows below the cultivable land. Thus, creating a difficulty to irrigate the land. Different types of pumps are being employed to pump water at a higher level according to the availability of resources. Among other, the electrically or fuel operated pumps are found to be more popular due to their high pumping capacity. Besides their capacities, they also have certain drawbacks. In the case of rural hilly communities of Nepal, the lack of developed electricity supply facilities and also the high operating and maintenance cost of these pumps are some major drawbacks. Considering these particular drawbacks, a water-powered Coil pump is suggested to be one of the better alternatives. Simply, coil pump is a setup of long pipe winded around a cylindrical drum half submerged in water, can be rotated by any mechanism, which on rotation enables water and air partially to gulp into the pipe and thus generate head at the outlet. It uses the principle of hydraulic pressure to lift the water.

On the riverside farms, water-powered Coil pumps can be employed for pumping. The pump does not require any external sources of energy as it can be rotated by utilizing the power of river water by attaching a propeller setup at one end of the pump shaft. Performance of two type of such water powered-coil pump, namely cylindrical and conical shaped pumps are studied, both experimentally as well as numerically. The first pump consists of a coil winded on the cylindrical surface whereas the second pump consists coil winded on the conical surface. Both pumps work with the same principle rather the differences are in their shape and construction. The purpose is to analyse the possibility of increasing the pumping head and discharge by changing the shape of the drum. For the experimental study, two coil pumps were constructed and tested in Sunkoshi River (a river located in Sindhuli, Nepal). Whereas, for the numerical study comparison was made between the performances of these conical shaped and cylindrical shaped coil pumps. With the aim of optimizing the performance, a numerical study was performed by varying the number of turns in the pump. Though the numerical setup had difficulty to replicate the actual experimental environment, similar trend in result was captured.

1.1. Theory

A coil pump is a kind of positive displacement pump that runs at low speed. This pump is first invented by H.A. Wirtz in 1746. A typical coil pump consists of a coiled tube, rotary coupling, structural support, and a driving system. The driving system can be a waterwheel, propeller, motor, or even human or animal. One end of the tube is connected to the scoop and another end is connected to a rotary coupling. Outside of that rotary coupling, a delivery pipe is connected [1–3]. The pump is used by partially submerged in water. When the pump is rotated, the rotational motion leads to an alternate intake of water and air in sequential order. Every coil has alternate air and water column as shown in Figure 1. These columns of air and water result in the action of pumping [4].

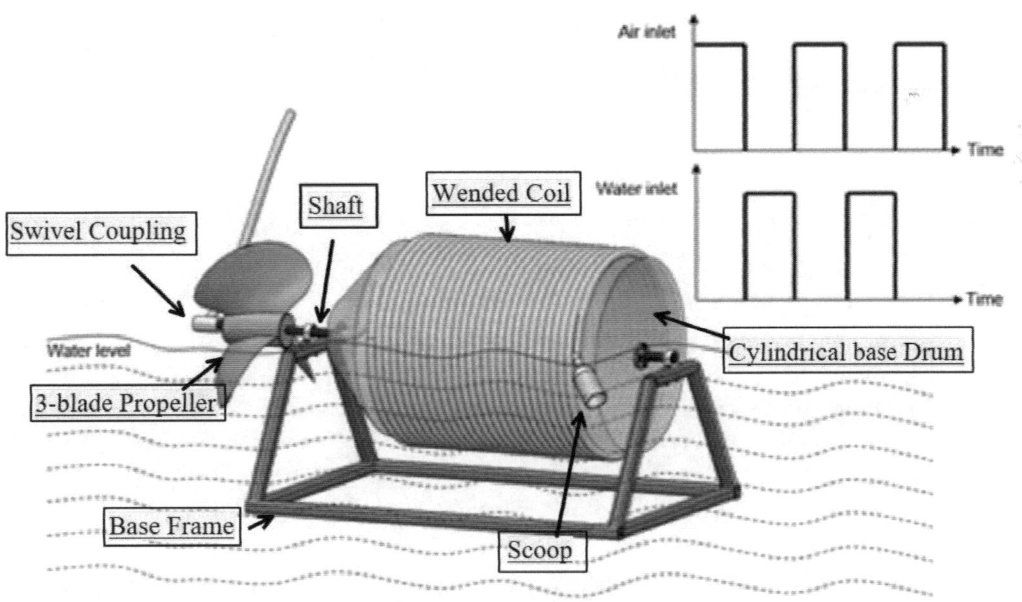

Figure 1. Parts of a Coil pump and illustration of the simultaneous water and air inlet as per the actual pump.

The pump revolution causes the forward motion of water and air towards the delivery pipe. This forward force develops pressure in each turn of the coil [5, 6]. The total pressure head produced at the pump is the sum of all the pressure created at each coil. The pressure at each coil is governed by the compression of each air column. This contraction is a polytropic process with a coefficient of 1.15 [7, 8].

$$H_A {L_A}^{1.15} = H_n {L_{An}}^{1.15} \tag{1}$$

The water flow rate is governed by the average amount of water that the coil inlet collects in each revolution. With no dynamic loss, the average water flow rate will be [7]

$$\mathrm{Q} = 2 \times \mathrm{R} \times \mathrm{A} \times \mathrm{N} \times \cos^{-1}(\frac{R-Di}{R}) \tag{2}$$

In the pump, both head and discharge are governed by fluid and system variables. Fluid properties include density, viscosity, surface tension, and internal roughness of pipe whereas system variables include diameter, pitch, rotational speed, and submergence of the pump [9, 10]. The static head of the coil pump increases with the increase in number of turns of the coil [4, 9]. It changes slightly with the changes in rotational speed [11]. It was found that at submergence 0% and 100%, the static head is zero. It increases as the submergence ratio increases from 0 to 15%, constant to 75%, and decreases afterward [4, 5]. Discharge of the pump is also found to be directly dependent on the diameter of the tube, diameter of coil, submergence, and speed of rotation [10, 11]. Discharge is slightly dependent on the number of turns in the coil. When the depth of immersion of the pump is increased, a higher amount of water and less amount of air enters the pump. This results in higher discharge and lower head.

The flow rate of the pump at the outlet depends on the amount of water collected by the scoop at every revolution of the coil. When no water is lost during the pumping action, the discharge in a revolution is the amount of water collected in the scoop in that revolution. The average flow rate is [7]:

$$Q = L_{wi} \times A \times N \tag{3}$$

With no dynamic loss [7],

$$L_{wi} = \theta \times R \tag{4}$$

Where, θ is angle subtended by water plug in intake coil and given by [7]

$$\theta = 2\cos^{-1}(R - D_i R) \tag{5}$$

Where, L_{wi} is length of water plug length in intake coil, R is the radius of curvature at the intake coil, A is cross-sectional area of the tube, N is the revolution per second, D_i is the depth of immersion.
The static head of the pump depends on the number of coils and submergence ratio. As the number of turns of the coil increases, the number of manometers is also increased which results in an increase in the static head [4, 12]. If the speed of rotation increases, discharge increases but the volume delivered per revolution decreases[13]. An increase in rotational speed increases the effect of centrifugal force on the air-water plug.

In order to increase efficiency and performance, a scoop is used. It is generally used when the depth of immersion is smaller as compared to the radius of the coil. Scoop is made with a larger diameter tube generally about 100-120% of outer coil volume [14, 15]. The pump performance is also affected by rotation. At higher rotational speed, centripetal force creates vortices, and pressure gradient and breaks essential airlock [1, 16–18]. This might result in a decrease in pressure and lift.

2. Materials and Methods

2.1 Experimental Setup

The performance of cylindrical and conical coil pumps in terms of head and discharge was to be studied. For this both the pumps were fabricated with the drum diameter of 700 mm (which, in case of conical pump, is the largest diameter). Double layered one-inch pipe of 80 meters length was winded around each base drum. A scoop was fixed at one end and T-joint was connected to other end, which was then connected to the extension pipe through the swivel coupling. The swivel coupling allows one side of the coupling to rotate while other side being stationary. The shaft of the coil pump needs to be rotating and the extension pipe needs to be stationary, this can be possible by the use of swivel coupling. A propeller with three blades was fixed in the shaft, which goes all the way through the drum and is connected to the swivel coupling. The parts, their materials and the quantity for both the pumps are listed in the Table 1. The parts installed in a Coil pump setup is pointed in Figure 1.

Table 1: Fabrication details of cylindrical and conical pumps

Parts	**Materials / Quantity**	
	Cylindrical Pump	*Conical Pump*
Drum	MS sheet	Glass fiber
Wended Pipe	HDPE (80 meters)	HDPE (80 meters)
Propeller	Aluminum	Glass fiber
Frame	MS pipe	MS pipe
Shaft	GI pipe (2.25 meters)	GI pipe (2.25 meters)
Bearings	3 pieces	3 pieces
Swivel coupling	1 piece	1 piece

The images Figure 2 (a) and (b) are actual coil pump fabricated for testing. The setup could be optimized but was constrained by the availability of material during the time of fabrication. Both the pumps were installed in Sunkoshi River, Sindhuli for the testing. The pump was half immersed in the river. The average flow rate at the outlet of the pump was measured for the different height differences.

(a)

(b)

Figure 2. Working model of (a) conical pump, (b) cylindrical pump

2.2 Numerical Setup

Numerical analysis was carried out using OpenFOAM, which is an open-source CFD tool. The geometry and boundary conditions were defined replicating the experimental work. However, because of mesh generation and other complication, the simulation was performed in a simplified domain as shown in Figure 3. The domain consists of inlet, outlet and wall boundary condition. Water and air were injected into the domain via inlet patch. Outlet patch was defined to have atmospheric pressure, while the wall was considered to be no-slip walls. The boundary conditions applied can be found in the Table1.

The pump was designed for 15 m head (excluding system loss) with 20 rpm to discharge of 15,000 liters per day. To replicate the design conditions the domain was rotated along the axis of the parent geometry (cylinder and cone) at 2.0944 rad/s or 20rpm. At the inlet patch the time of injection of water into the domain was calculated using the rpm of the pump as reference. The pump was submerged in water; thus, water enters into the domain only during the half phase of a rotation. Taking the rotational speed to be 20rpm, it takes 3 seconds for a single rotation. Hence, water gets injected into the domain for 1.5 seconds and air for next 1.5seconds. This pattern of injection was defined in alpha.water file of 0 folder in OpenFOAM.

Table 2. Patch names and their corresponding boundary conditions.

Patches	P_rgh	U	Alpha.water
Inlet	*zeroGradient*	*volumetricFlowRate constant 0.000371*	*uniformFixedValue, uniformValue table ()**
Outlet	*fixedValue, uniform 0*	*zeroGradient*	*inletOutlet, uniform 0*
Wall	*zeroGradient*	*noSlipS*	*zeroGradient*

* *table was defined so as to inject water in interval of 1.5s*

The multi-layered coil pump setup was used in experimental study. Due to complication in generation of mesh in such domain, single layered coil pump setup was used in the study, while approach was made to study the effect in having different length of the coil.

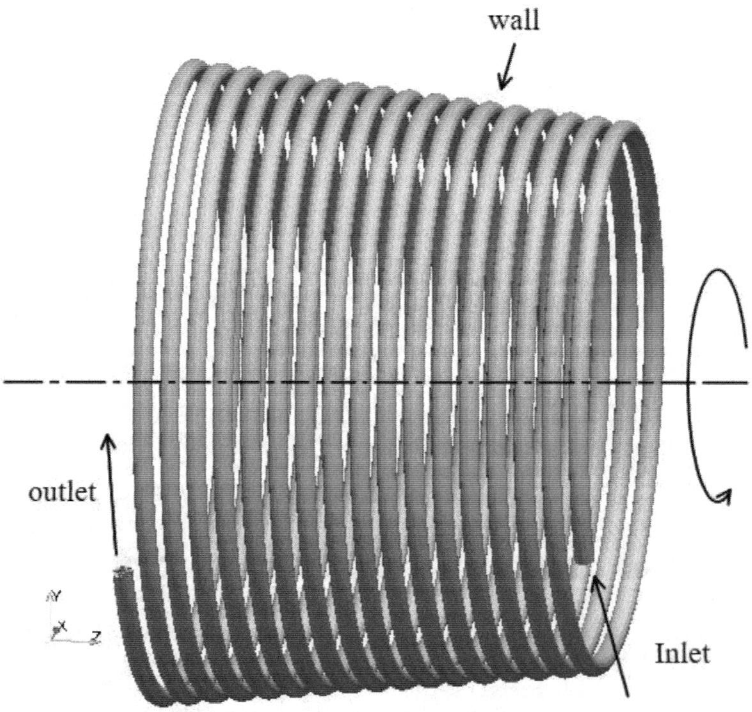

Figure 3. Boundaries in the computational fluid domain and their patch names.

The fluid domain for both cylindrical and conical coil pump were created in SolidWorks. Further two cases were studied for coil with 15 rounds and 7 rounds of coil. The nomenclature of the cases is done as L1 for 15 turns, and L2 for 7 turns. The Figure 4 shows the meshes of the created fluid domain. The meshing was done in ANSYS meshing tool. The meshing details for each case is shown in Table 3.

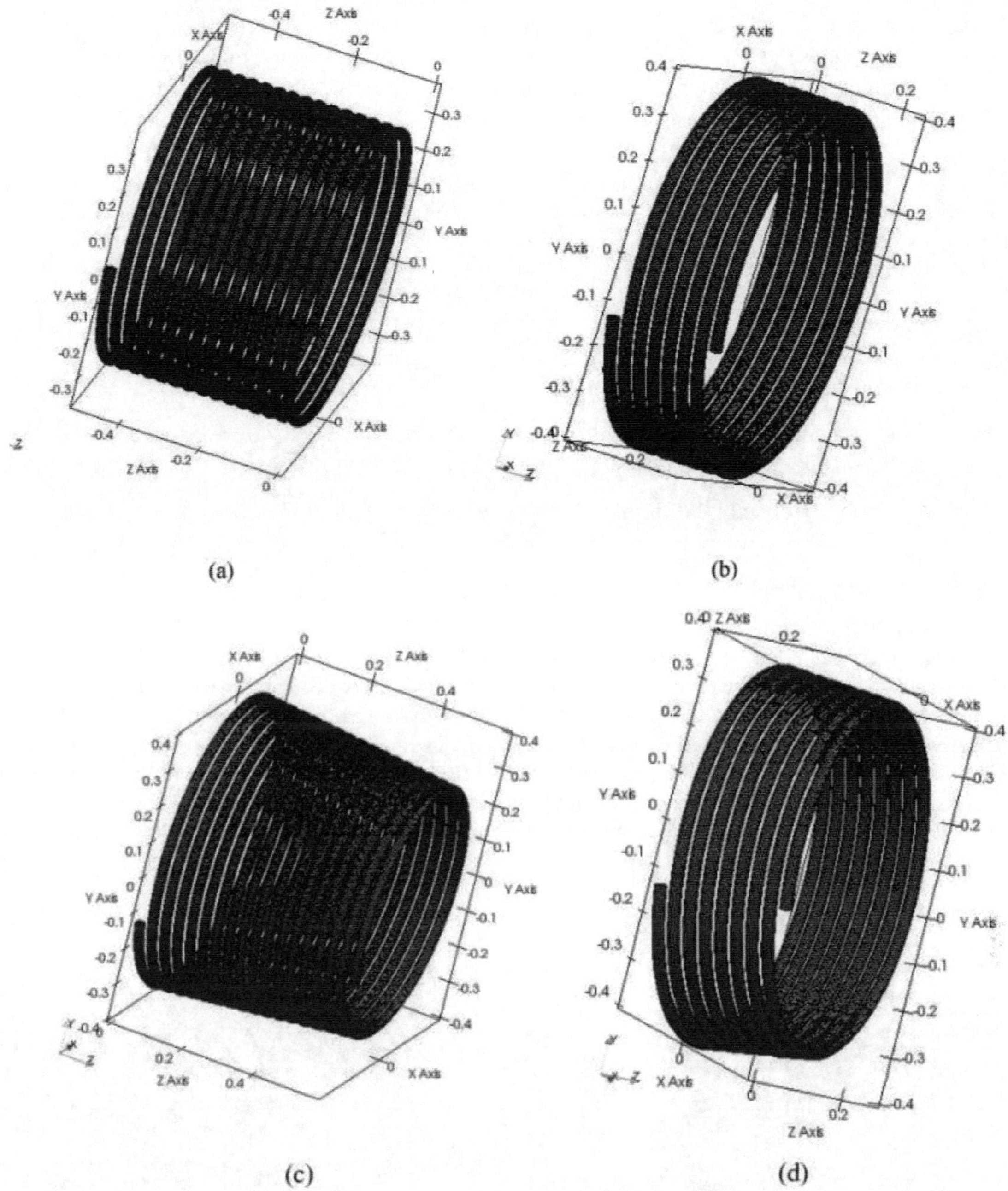

Figure 4. Complete domain mesh for (a) cylindrical coil with 15 turns (b) cylindrical coil with 7 turns (c) conical coil with 15 turns (d) conical coil with 7 turns

Table 3. Mesh statistic details

Case	Number of elements in Mesh
Cylindrical L1 (15turns)	37424
Cylindrical L2 (7 turns)	22082
Conical L1 (15 turns)	19218
Conical L2 (7 turns)	15631

The domain has air and water phases at a same time so, interFoam solver was used. The interFoam uses Volume of Fluids (VOF) method to track the multi-fluid interaction and multidimensional universal limiter for explicit solution (MULES) method to maintain boundedness of phase fraction independent of meshing and selected numerical schemes [19]. In order to simplify the simulation, flow was considered to be laminar. The simulation was carried out for 71 seconds in 12 processing cores for each case, after which the pattern of result at the outlet was consistent. The simulations were performed at Supercomputer Centre, Kathmandu University. The computational time taken for each case is shown in the Table 4.

Table 4. Computational time taken for simulating each case

Cases	Computational time (s)
Cylindrical L1 (15turns)	179,795
Cylindrical L2 (7 turns)	67,912
Conical L1 (15 turns)	111,042
Conical L2 (7 turns)	36,962

3 Results and Discussions

3.1 Experimental Results

The pumps were installed in Sunkoshi river of Sindhuli district, Nepal. **Figure 5** (a) and (b) are the images from the testing of the pumps in the river. The flow velocity of the river was found to be 0.9 m/s, in which the conical and cylindrical pump rotated with the rotational speed of 11 and 17 rpm respectively. The discharge was measured in the same varying condition, as there was no any measure to maintain the rotation during the study. The average flow rate at the outlet of the pump was measured for the different height differences. The maximum discharge of 12000 and 8800 liters per day was obtained for conical and cylindrical pump respectively, while the maximum head was 6.5 meters and 11 meters.

(a)

(b)

Figure 5. Installation of two pumps (a: Conical pump, b: Cylindrical) in Sunkoshi river, Sindhuli

For both pumps, it was found that the discharge decreases with the increasing head or lift. The discharging capacity of the conical coil pump was observed to be higher than that the of cylindrical coil pump for zero lift whereas the maximum lifting capacity of the cylindrical coil pump was observed to be higher than that of the conical coil pump. The obtained results are shown in the performance graphs, Figure 6. The testing conditions, flow velocity of the river and the rotational speed of the impeller were less than the designed conditions that resulted in the lower lifting and discharging capacities of both the pumps.

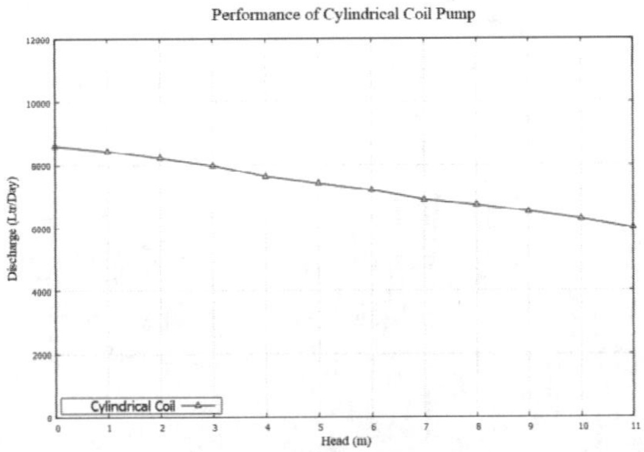

(a)

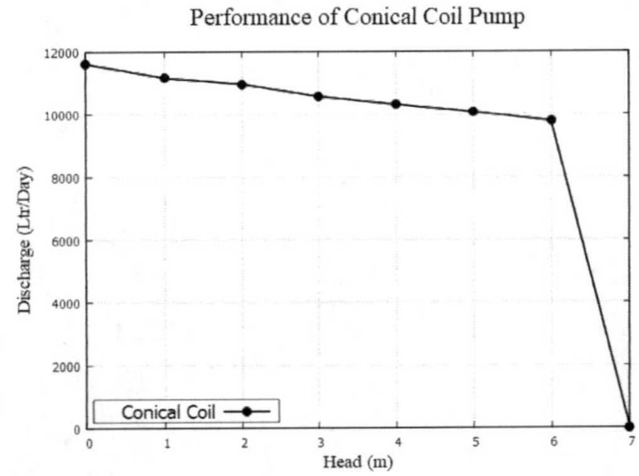

(b)

Figure 6. Experimental performance of (a) Cylindrical and (b) Conical Coil pump

3.2 Numerical Results

Numerical results were extracted using built in function objects in OpenFOAM. Water volume and Pressure at the outlet region were calculated for all four cases and plotted over time. The results were visualized using ParaView and graphs were plotted using gnuplot.

3.2.1 Water Volume at Outlet

The flow of water from the outlet was found to be resembling the experimental case. As per the literatures, the air and water come out from the outlet in an alternate cycle. The pattern of water flow from the outlet patch is shown in **Figure 7**. A comparison was made in between the pattern of water volume fraction at outlet of cylindrical and conical setup. A distinctive pattern that can be concluded from the graph is that, the water volume in the case of conical pump is better than in cylindrical pump.

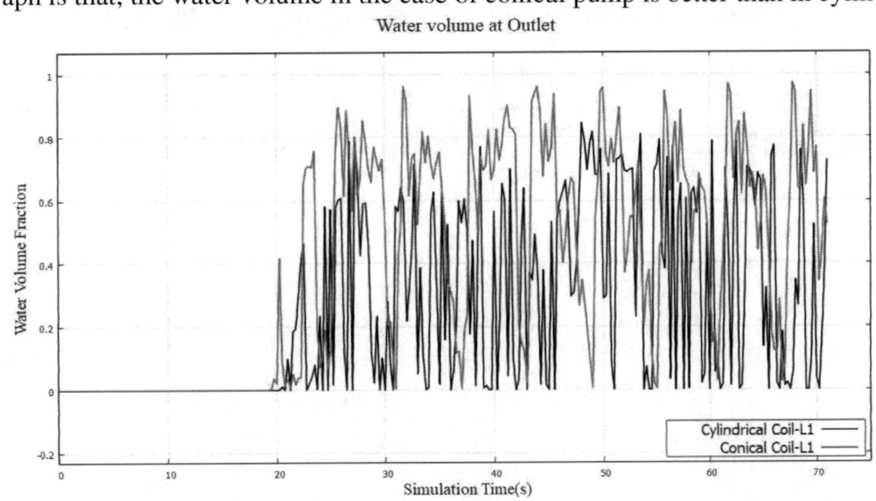

Figure 7. Water volume fraction comparison of cylindrical and conical at outlet

3.2.2 Pressure at Outlet

The pressure at outlet of the coil was calculated using built in function of OpenFOAM, *surfaceFieldValue*. This function provides the area average value of a parameter at certain patch, in this case Pressure at outlet patch. The calculated pressure values were plotted over time. Delivery of water at outlet was seen in similar pattern as in experiment. Both positive and negative pressure values were obtained in random cyclic manner for all the four cases. The positive pressure in the graph was obtained when there is water discharge at outlet. After the completion of discharge cycle, water near the outlet is pulled back to its lowest level due to gravity causing suction phenomenon. This phenomenon can be the reason behind the occurrence of negative pressure at outlet patch.

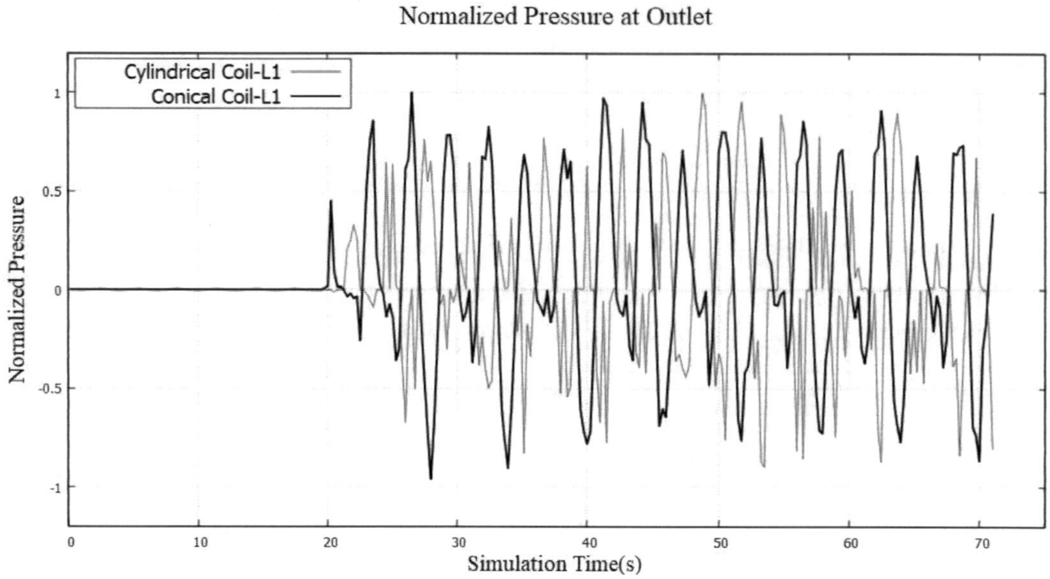

Figure 8. Pressure comparison at Outlet in cylindrical and conical coils

The graphs as show in Figure 8 and Figure 9 are normalized against the maximum pressure. From graph in Figure 7, it is found that the head obtained in cylindrical coil is better than that in conical coil for same number of turns. The average pressure deviation between conical and cylindrical coil was calculated to be 2.52% for 15 turns and 10.88% for 7 turns. This can be noted due to the uneven length of the tube winding because of the geometrical shape, i.e., tube length in cylindrical coil is more than that in conical, thus causing increase in pressure at outlet in cylindrical setup.

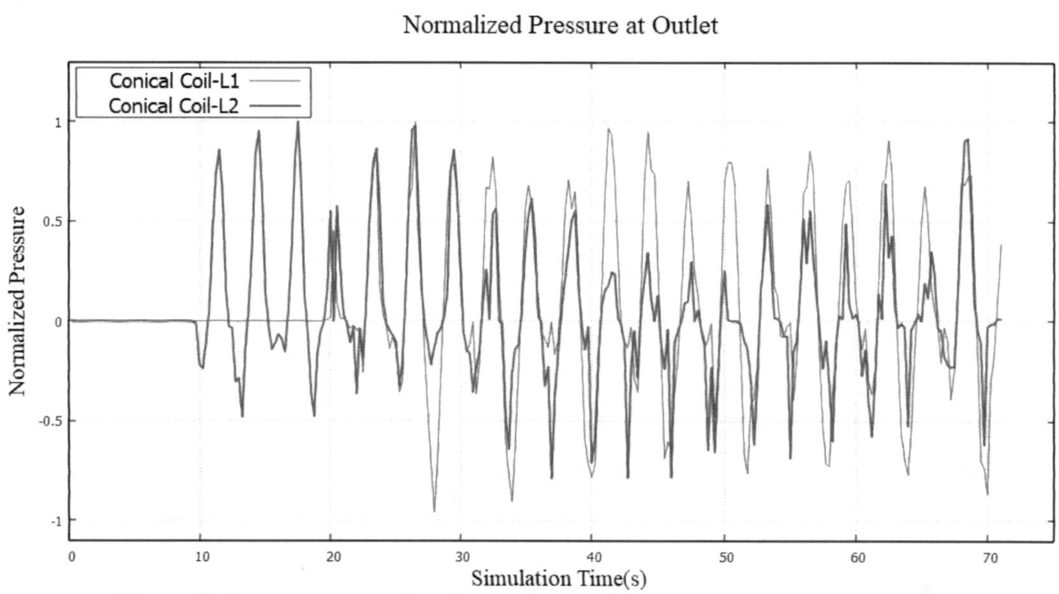

Figure 9. Pressure comparison between conical coil with 15 turns and 7 turns

Another comparison was made in between conical coils with 15 turns and 7 turns, as shown in Figure 9. It is seen that the pressure increases as we increase the number of turns. The deviation while increasing the number of turns, from 7 turns to 15 turns, in cylindrical surface was found to be 34.91%, while for the case of conical it was found to be 42.55%.

4 Conclusion

Installation of water powered coil pump can help in solving the problem of irrigation in rural Nepal. The Coil pump can be manufactured and maintained in local community. The pump being water powered, it can operate without human assistance. The pump is suitable for lifting water for low height difference and can be a good option for rural communities of Nepal.

Experimental testing of two type of coil pump, cylindrical and conical, was performed in nearby river (Sunkoshi) of Nepal. Both the pumps had equal length of pipe windings but the rotational speed was varying due to the flow condition of river. The maximum discharge of 11500 liters/day from conical pump was measured, which decreased while increasing the head. In case of cylindrical pump, the maximum discharge of 8600 liters/day was measured, which was lower than conical. However, the pumping head was found to be maintained even after 11m, with the discharge of about 6000 liters/day at 11m head.

The numerical study was performed for visualizing the flow in coil pump. Four different cases were studied, in which the number of turns and shape of coil were varied. Geometry of each case was created using SolidWorks and meshing was done using ANSYS meshing tool. OpenFOAM was used for numerical study. Since, there is interaction of two phases (water and air) in the domain, interFoam solver was used. For simplifying the study, laminar turbulence model was employed. Water and air were injected into the domain from the inlet patch at calculated time rate. The deviation in average pressure in cylindrical and conical pump from numerical study was calculated as 2.52% while using 15 turns. The outlet pressure was found to be increasing by 35% while increasing the number of turns from 7 to 15 in cylindrical and 42.5% in case of conical coil pump.

Comparison was done between the performance of cylindrical and conical shaped coils both experimentally and numerically. Due to constrain of testing environment and computation environment, the experimental setup could not be similar to numerical study. However, the flow pattern as obtained in numerical study was similar to that in experiment. It was found that the cylindrical pump was better for higher head, whereas the conical pump had shown better results for higher discharge. The results from OpenFOAM were found to be incoherent to the experimental results and the study also helped to understand and predict the flow pattern in the coil pump. Further work can be carried out for optimizing the performance of the pump. The areas that can be focused for optimizing are the number of turns in pump, and use of multiple inlets.

5 Acknowledgements

This work is supported by **Nepal Technology Innovation Center (NTIC)** under the auspices of **Korean International Cooperation Agency (KOICA)**. The computational support for the numerical analysis is provided by the **Kathmandu University Supercomputer Centre**.

6 References

[1] Quiroga J, Tschiersch K & Bohórquez O, Coil pump design as an object of meaningful learning, *J. Phys. Conf. Ser.*, 1161 1 (2019) 012027.

[2] Ewbank T, A descriptive and historical account of hydraulic and other machines for raising water, Scribner, Armstrong (1876).

[3] Naegel L C A, The hydrostatic spiral pump: Design, construction and field tests of locally-developed spiral pumps, July (1998).

[4] Kassab S Z, Naby A A A, Sayed E & Basier I A, Coil Pump Performance Under Variable Operating Conditions, (2005) 655.

[5] Patil N R, Gaikwad S R, Navale R A & Sonawane D S, Design, Manufacturing and Performance Analysis of Spiral Coil Pump, *Appl. Mech. Mater.*, 446–447 (2014) 549.

[6] Dangwal K K & Aggarwal M, Irrigation using Natural Energy Sources, *Int. J. Recent Technol. Eng.*, 3 (2019) 2277.

[7] Mortimer G H & Annable R, The coil pump-theory and practice, *J. Hydraul. Res.*, 22 1 (1984) 9.

[8] Mortimer G., *The coil pumps -Doctoral Thesis*, Loughborough University of Technology, UK (1988).

[9] Maeda T, Design and parametric performance of a rotating : helical coil pump, *Fluid Power*, (1993) 262.

[10] Modi V J, On the Development of a Wind Energy-Based Efficient Irrigation System, *SAE Tech. Pap.*, (1999).

[11] Ramli M I, Basar M F & Razik N H A, Natural Energy Water Pump: Revisit the Water Sling Pump, *Int. J. Innov. Technol. Explor. Eng.*, 3 2 (2013) 188.

[12] Modi V J & Nourbaksh A, Design and parametric performance of a rotating helical coil pump, *Proc. JFPS Int. Symp. Fluid Power, Tokyo, 1993*, 1993 2 (1993) 249.

[13] Hilton D J, Further development of the inclined coil pump, *Waterlines*, 8 2 (1989) 20.

[14] Naegel L, Real J & Mazaredo A M, Designing a spiral pump for irrigation, *Waterlines*, 10 2 (1991) 30.

[15] Chogale G, Kolhe K, Shrotriya V, Talgaonkar A & Pokharkar T, Review Paper on Design and Development of Manually Operated Wirtz Pump, *Int. J. Sci. Res. Dev.*, 5 11 (2018) 749.

[16] Tailer P, The Spiral Pump. A High Lift, Slow Turning Pump, (1986) p. 35.

[17] Zhu G, Yang X, Jiang S & Zhu H, Intermittent gas-liquid two-phase flow in helically coiled tubes, *Int. J. Multiph. Flow*, 116 (2019) 113.

[18] Murai Y, Yoshikawa S, Toda S I, Ishikawa M A & Yamamoto F, Structure of air-water two-phase flow in helically coiled tubes, *Nucl. Eng. Des.*, 236 1 (2006) 94.

[19] OpenFOAM, OpenFOAM User Guide version 8, July (2020) U2.

Assessing the performance of a demonstrative hydrogen fuel cell power train in the chassis of an internal combustion engine vehicle

Y. Shrestha[1], A. Subedi[1], M. Sapkota[1], P. Paudel[1], N. K. Yadav[1], B. S. Thapa[1*]

[1]Green Hydrogen Lab, Department of Mechanical Engineering, Kathmandu University, Dhulikhel, Nepal
*Corresponding address: bst@ku.edu.np

Abstract. In the present day in the global context, approximately 2 billion internal combustion engines (ICEs) are estimated to be in use. Nearly 7.3 billion metric tons of carbon dioxide (CO2) were released into the atmosphere as per the survey on the transportation sector in 2020, making up 41% of all emissions. To solve pollution challenges related to ICEs, recent advancements of alternatives to ICEs have pushed the globe to establish clean energy sources such as hydrogen fuel cells. The conversion of an ICE vehicle to a fuel cell vehicle is able to add further support to its development besides the 30% emission reduction since the old ICE vehicle would not be discarded. FCEVs are seen as a true replacement between BEVs and ICEs since more energy can be stored in less space as compared to BEVs. The objective of this paper is to analyse the energy consumption of a fuel cell-driven motor in the internal combustion engine vehicle chassis of Maruti 800, through mathematical modelling, then compare the specifications with that of an ICE-driven system. The analysis provides details on the optimization of drive system components to maximise the fuel cell energy consumption efficiency based on fuel cell power source requirements for a range of velocities and road gradients. The modifiable specifications of drive system components in the mathematical model ensure reproducibility for any drive cycles and system scenarios.

Keywords: Hydrogen fuel cell, green hydrogen, vehicle modelling, mathematical modelling

1. Introduction

The current world is dominated by vehicles running on Internal combustion engines. As per the data, in 2019, 90 percent of the total automobile sales in the world were of IC engine vehicles [1]. Fossil fuel is not a sustainable energy source, yet it is one of the primary sources of energy production for all nations. To supply the energy demands of the more rapidly increasing global population, it is essential to upgrade to an alternative, sustainable energy source with no adverse consequences to the environment. This effect on the environment has raised many concerns around the globe and much research is being done to decrease the reliance on fossil fuel as a source of energy. Alternatives including Battery-powered Electric Vehicles (BEV) and Fuel Cell Vehicles (FCV), which can replace ICE with ones that are more environmentally friendly, are being tested because the bulk of today's cars use fossil fuels [2].

In a manner similar to how an electrochemical battery works, a fuel cell transforms the chemical energy of hydrogen and air into heat and direct current electrical energy. Unlike a battery which stores energy, by providing fuel to the fuel cell, the converted energy is supplied continuously to provide power [3]. The majority of fuel cell vehicles feature a hybrid setup (Figure 1) that combines a high voltage battery and a fuel cell stack to harvest braking energy for improved acceleration and increased range [4].

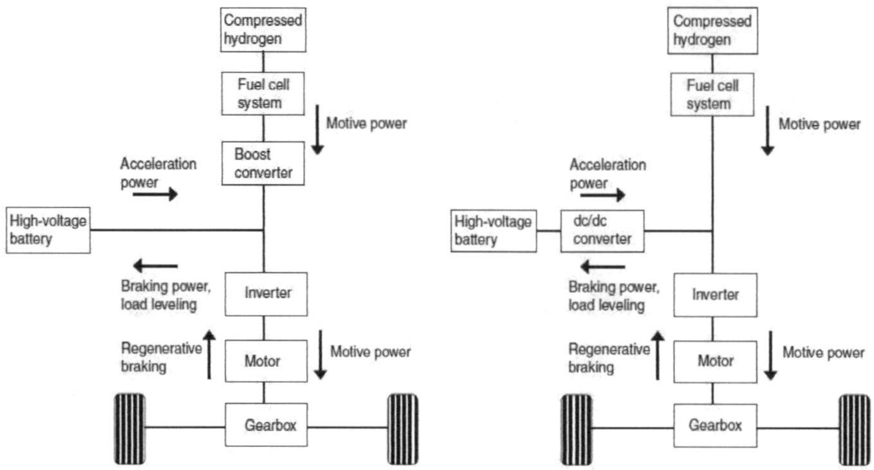

Figure 1. Automobile propulsion system [4]

2. Vehicle load forces

The primary load forces to be taken into consideration which resists a vehicle's forward motion is the tractive force which is the combination of aerodynamic drag, rolling resistance, inertial resistance and gradient forces as depicted (Figure 2). The tire material, tire structure and tire temperature all affect the rolling resistance coefficient. The shape of the vehicle determines the drag coefficient [5].

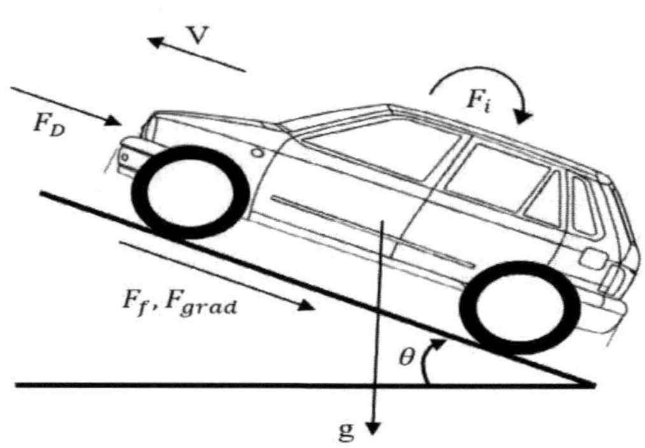

Figure 2. Vehicle load forces

3. Past research work

The review of the paper titled "Parameters Matching for Electric Vehicle Conversion" provides insight into the process of modular conversion of conventional IC vehicles to electric vehicles [6]. The concept of conversion to a specific car is discussed in the paper, along with calculations for the electric motor and battery and access to the vehicle's original specifications. The conclusion of the paper is defined as the parameter optimization requirements for the conversion process to achieve the required speed, driving range and road gradient.

The analysis by F. Barbir in the paper titled "Fuel Cell Vehicle" details the working parts of a hybrid fuel cell/ battery electric vehicle [7]. Similarly, calculations are made for the fuel consumption and range and plotted for a certain constant speed. The fuel cell stack, drive motor/ controller and air compressor are identified as the factors influencing performance based on the data from the graph. The article's conclusion outlines the process of modifying an existing composite two-seater car by adding 15 kW fuel stacks and the necessary tools to enable it to go at a top speed of 95 km/hr while weighing 1360 kg and having a range of 100 km. The particular fuel cell stack specifications are also provided for the ease of readers.

The review of the thesis "Design and Control of Hydrogen Fuel Cell Vehicle" by Joseph Kenneth Smithson Bell provides information on the conversion of a 1969 Volkswagen Beetle into a PEMFC hybrid vehicle powered by hydrogen [8]. The required calculation for the power required at different stages of the vehicle is simulated in MATLAB software along with the hydrogen consumption based on the data from the general drive cycle velocity, which along with the drive train component efficiencies provides the accurate fuel cell power calculation.

The fuel cell hybridization component is presented in a fuel cell electric vehicle review work by Mustafa Inci, et al. [9]. Batteries, ultracapacitors, solar panels and combinations of auxiliary systems are some examples of fuel cells that can be hybridized with.

4. Fuel cell vehicle powertrain

In the powertrain of a fuel cell hybrid system (Figure 3), four different energy flows are shown, a direct energy flow from the fuel cell to load, a charging energy flow from the fuel cell to the energy storage system, an energy flow of regenerative braking, and finally an energy flow from the energy storage system to the load [10].

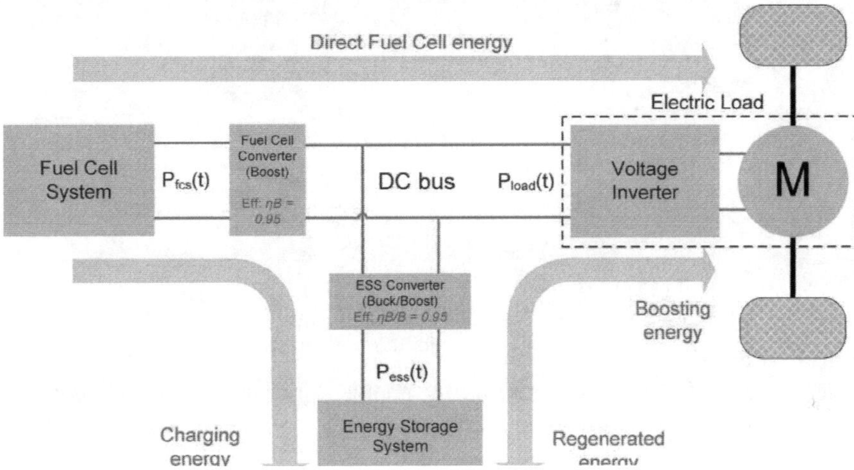

Figure 3. Powertrain of a fuel cell hybrid system [10]

5. Topology of electrical system

Four elements are taken into consideration when analysing topologies: load characteristics, energy recovery potential, storage system type and storage location. The flowchart (Figure 4) defines how the consideration process is carried out [10].

Based on the flowchart, the selected topology for the project can be broken down into: the type of load is DC; hence the DC/DC converter is present before the load. No regenerative braking is applied to imply the use of a unidirectional DC/DC converter. The type of ESS used is a battery bank only and it is placed in parallel with the fuel cell after the converter.

In recent years, a variety of mathematical models have been developed to simulate the drive systems of automobiles. The published work of several authors has modelled internal combustion engine, fuel cell engine, battery electric and hybrid drive systems to generate criteria of vehicle modelling, the present study aims to study the case of power source requirements of fuel cell system if installed in an existing IC engine car chassis [8], [11], [12]. We simulated a standard driving cycle through our mathematical model and analyzed the necessary power and energy requirements for the selected scenario.

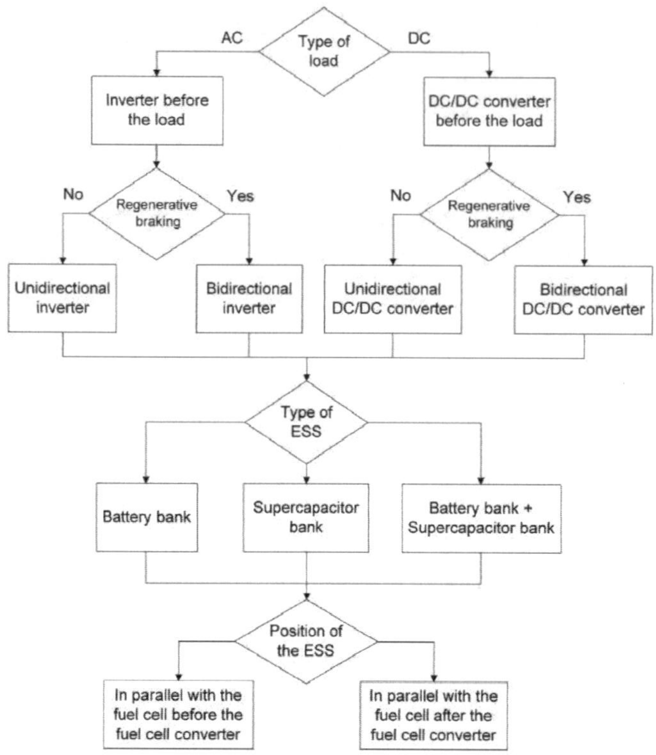

Figure 4. Topology selection flowchart [10]

6. Materials and methods

The mathematical model is created in Simulink for analysis (Figure 5). The standard data of class 3-WLTC driving cycle is the base input with a saturated maximum value for velocity to match out targeted speed of 50 kmph [13]. The acceleration reaches up to 1.67 m/s^2 and road gradients of 0°, 5°, 10°, 15° and 20° are considered. The vehicle specifications of Maruti Suzuki 800 Alto Car are referred for information on static mass, wheel radius, frontal area and gear ratios specific to the car [14]. The applied constants for the considered scenario (Table 1) and the standard formulae (Equations 1-15, Table 2) for vehicle modelling are utilised throughout the model [8], [9], [15]–[21].

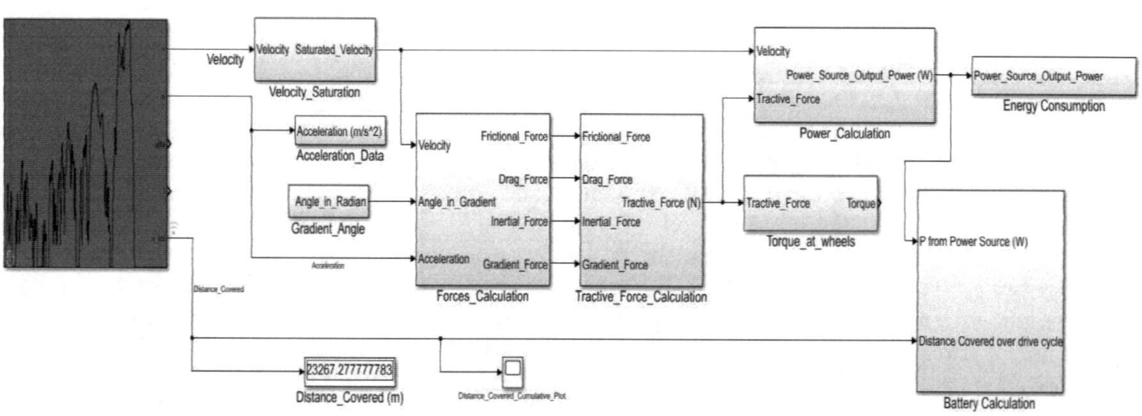

Figure 5. Mathematical Model of Fuel Cell Drive System

Table 1. Assumed Parameters and References

Assumed Parameters	Values	Units
Air Density (ρ)	1.2	kg/m^3
Coefficient of drag resistance (C_d)	0.35	
Coefficient of rolling resistance (C_r)	0.01	
Acceleration due to gravity (g)	9.8	m/s^2
Electric motor efficiency (η_m)	0.9	
Fuel cell Efficiency (η_f)	0.6	
Motor controller efficiency	0.95	
DC-DC controller efficiency	0.9	
Lower Heating Value of Hydrogen (LHV)	120	kJ/gm
Wind Speed (v_a)	2.22	m/s
Radius of wheel	0.2684	m
Frontal Area (A_f)	1.93	m^2
Transmission efficiency(η_t)	0.85	
Total static mass (m_t)	800	kg

Table 2. Equations used for mathematical model design

Equations in successive sequence	
$\text{Drag Force } (F_d) = \frac{1}{2} \times A_f \times \rho \times C_d \times (v + v_a)^{\wedge}2$	(1)
$\text{Frictional Force } (F_f) = m \times g \times \cos\theta \times C_r$	(2)
$\text{Inertial Resistance } (F_i) = m_{eff} \times a = 1.035 \times m_t \times a$	(3)
$\text{Gradient Force } \left(F_{grad}\right) = m_t \times g \times \sin\theta$	(4)
$\text{Total Tractive Force Required } (F) = F_d + F_f + F_i + F_{grad}$	(5)
$\text{Max Power Demand } (P_d) = F \times v$	(6)
$\text{Min. Power Rating of power source } \left(P_p\right) = \frac{P_d}{\eta_m \times \eta_t \times \eta_{DC-DC\ Converter} \times \eta_{Controller}}$	(7)
$\text{Power Delivered by motor } (P_m) = \frac{P_d}{\eta_t}$	(8)
$\text{Power Delivered to motor } (P_{deL}) = \frac{P_m}{\eta_m}$	(9)
$\text{Average Power Output per Drive Cycle Distance} = \frac{\text{Average Power Output (W)}}{\text{Drive Cycle Range (km)}}$	(10)

$$\text{Required Power Output} = 80 \times \text{Average Power Output per Drive Cycle Distance} \left(\frac{\text{W}}{\text{km}}\right) \tag{11}$$

$$\text{Current Rating (A)} = \frac{\text{Required Power Output (W)}}{\text{Voltage Rating (V)}} \tag{12}$$

$$\text{Power Source Charge Capacity (Ah)} = \frac{\text{Energy Output Demand from Power Source (W − h)}}{\text{Voltage Rating (V)}} \tag{13}$$

$$\text{Energy Consumed by Power Source (Joules)} = \text{Energy consumed by Power Source (W − s)} = \text{Energy consumed by Power Source (kW − h)} * 3.6 * 10^6 \tag{14}$$

$$\text{Amount of Hydrogen Required (gm)} = \frac{\text{Energy Consumed by Power Source (Joules)}}{\text{Lower Heating Value of Hydrogen } (\frac{\text{J}}{\text{g}})} \tag{15}$$

7. Results and discussion

7.1. Power source specification

The power required for a motor in a drive train is supplied from a power source, which is fuel cell in this case. Based on the power output graph (Figure 6) for 0° gradient, the maximum power output is 33.20 kW. Since real road drive conditions may provide a much more flexible power requirement, average power output over the graphs (Figure 6) is used to calculate the unit power required over the drive cycle distance. Since the drive cycle distance is only 23.267 km, a range of 80 km is assumed for better representation of our targeted vehicle specifications. Similarly, the cumulative energy demand (Figure 7) gained by integration of the power demand from power source provides the maximum energy demand by the drive system and its product with assumed voltage rating of power source, i.e. 250 Volt, provides the power source charge capacity as shown in power source specification result (Table 3).

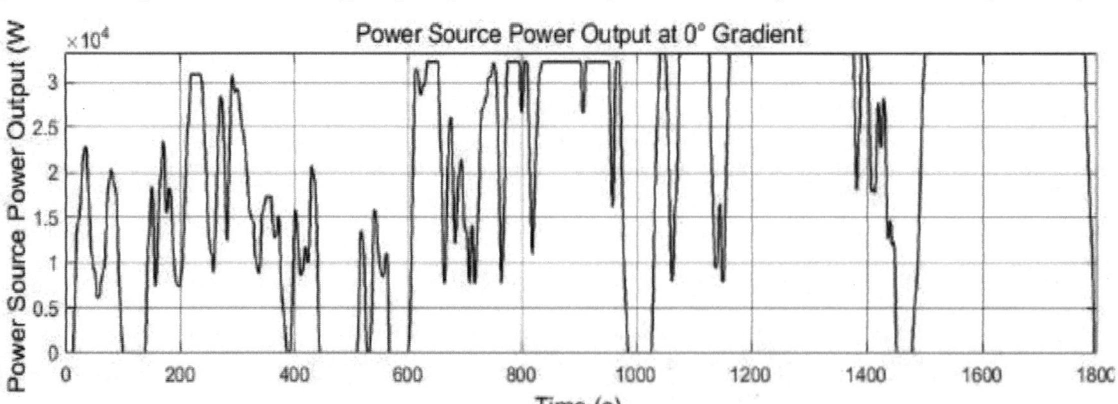

Figure 6. Power source power output simulation profiles

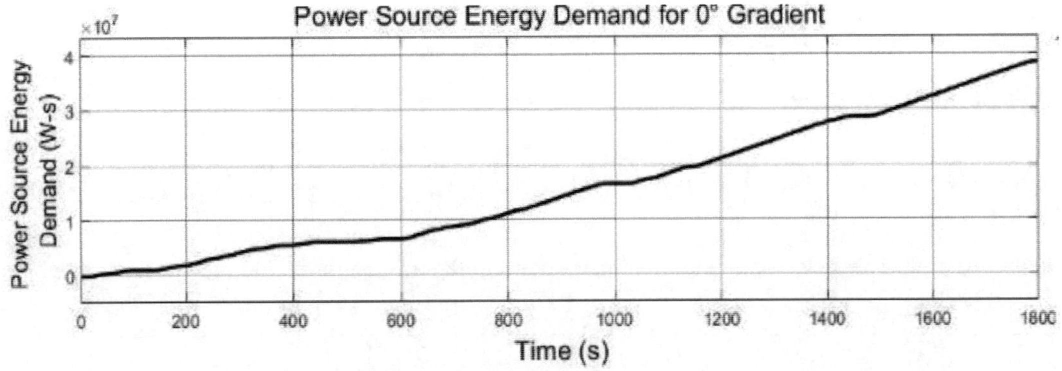

Figure 7. Power source energy demand simulation profiles

Table 3. Power source specification table

Calculated parameters for power source	Gradient (°)				
	0	5	10	15	20
Maximum power output (kW)	33.2	48.21	63.1	77.74	92.03
Average power output (kW)	21.41	31.36	41.21	50.91	60.38
Average power output per drive cycle distance (kW/km)	0.92	1.35	1.77	2.19	2.6
Ampere (A) if motor voltage is 250 V	132.8	192.84	252.4	310.96	368.12
Cumulative energy demand from power source (μW-s)	38.56	56.47	74.23	91.69	108.75
Energy demand from power source (kW-h)	10.71	15.69	20.62	25.47	30.21
Power source charge capacity (A-h)	148.78	217.87	286.37	353.76	419.54

7.2. Calculated amount of hydrogen required

The amount of energy required (Figure 8) to extract hydrogen from fuel is equal to the power source's highest energy consumption during the drive cycle. Therefore, the conversion of energy in terms of kW-h to Joules simplifies the derivation of hydrogen consumed calculation. The procedure uses hydrogen with a lower heating value of 120000 J/g. Based on the simulation result (Table 4), the acquired 0.53 kg usage of hydrogen denotes feasible calculation parameters.

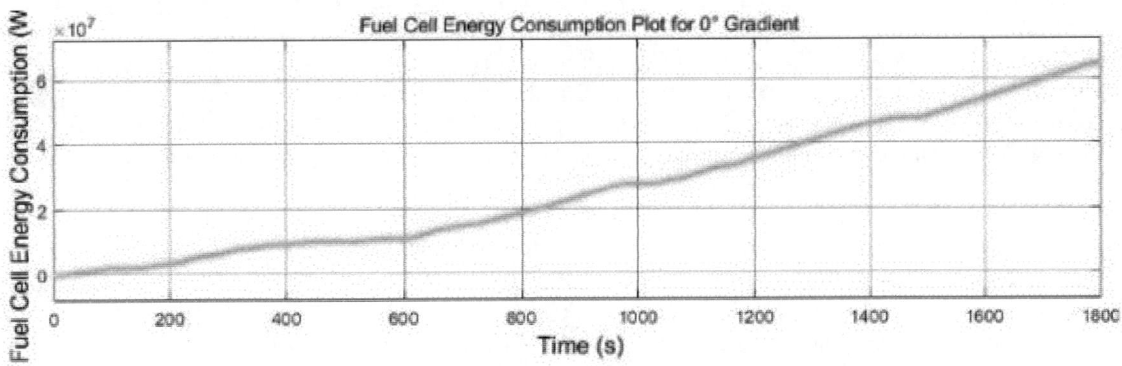

Figure 8. Energy consumed by fuel cell simulation profile

Table 4. Hydrogen amount calculation result

Calculated parameters	Gradient (°)				
	0	5	10	15	20
Maximum energy consumed by power source (kW-h)	17.85	26.14	34.36	42.45	50.35
Conversion of energy to joule (μJ)	64.27	94.12	123.71	152.83	181.24
Amount of hydrogen required (gm)	535.62	784.32	1030.93	1273.55	1510.35

7.3. Power source requirement variation with mass of vehicle

With considerations to the mass variations of vehicle from the base case i.e. 800 kg, with inclusions of heavier components/materials and additional passenger and luggage, three additional vehicle masses were considered, i.e. 950 kg, 1100 kg and 1250 kg. The average of added mass is considered to be 150 kg, which is the basis for the increment.

Based on the power output result (Table 5), the maximum power output in order to achieve 50 kmph at 0° gradient for vehicle mass of 950 kg is 52.44 kW. Considering the selection of 55 kW power output source, the vehicle would be able to achieve 40 kmph at 5°, 30 kmph at 10°, 20 kmph at 15° and 10 kmph at 0° road gradients.

The calculation for vehicle mass of 800 kg results in 33.2 kW power output to be the optimum requirement for power source. The calculation process is also repeated for vehicle mass of 1100 kg and 1250 kg, where, a power source of 65 kW and 70 kW is sufficient to achieve speed of 50, 40, 30, 20 and 10 kmph at 0°, 5°, 10°, 15° and 20° road gradients respectively.

Table 5. Power output variations (kW) for vehicle mass of 950 kg

Velocity (kmph)	Gradient (°)				
	0	5	10	15	20
50	52.44	69.66	86.74	103.54	119.94
40	41.39	55.17	68.83	82.27	95.39
30	30.7	41.03	51.28	61.36	71.2
20	20.29	27.18	34.01	40.73	47.29
10	10.08	13.53	16.94	20.23	23.58

7.4. Calculation reproducibility of the mathematical model

To verify equivalent input parameters, the model is placed through WLTC, NEDC and numerous standard driving cycles with saturated values. The power and energy outputs are nearly similar to that of the WLTC driving cycle. This result suggests the reproducible calculations for any drive cycle used.

7.5. Comparison of theoretical vehicle model with original vehicle

As per the specification of proposed FC Maruti 800, Hi-Tech petrol engine is replaced by hydrogen fuel cell as a power source and the simulation result is summarized (Table 6) [14]. The theoretical calculation suggests the maximum power of the converted vehicle increases from 37 bhp to 44.25 bhp. As the motor was changed, considering the torque multiplication in different gears in transmission system, the torque of the motor is calculated to be at a minimum of 136.59 Nm. Multiple Port Fuel Injection (MPFI) is replaced with pneumatic fuel system to go with the fuel intake system of fuel cell. The stock manual transmission system is fixed in single speed at second gear to be coupled with fuel cell. Front wheel drive is utilized with decreased gross weight from 1000 kg to 800 kg. For the demonstrative purpose, the vehicle's top speed in WLTC is saturated to 50 kmph.

Table 6. Comparison table between original Maruti 800 and proposed FCV model

Particulars	Specification of stock Maruti 800	Specifications of proposed FCV
Power Source	Hi-Tech Petrol Engine	Hydrogen Fuel Cell
Displacement (cc)	796	NA
Max Power	37 bhp	33 KW~ 44.25bhp
Max Torque	59 Nm	136.59 Nm (Motor)
No. of cylinder	3	NA
Fuel Supply System	MPFI	Pneumatic Fuel System
Transmission Type	Manual	Single speed fixed transmission
Drive Type	FWD	FWD
Fuel Type	Petrol	Green Hydrogen
Top Speed (Kmph)	144 kmph	50 kmph (capped for simulation)
Gross Weight	1000 kg	800 kg (base case)

8. Conclusion

A mathematical model for the proposed fuel cell driven car based on standard drive cycle data of WLTC and vehicle specifications for the car model Maruti 800 is designed to calculate the power requirements, energy consumption and modified specifications. The result demonstrates the comparative outcome between the original and modified vehicle. The proposed vehicle should theoretically be able to attain 50 kmph velocity in ideal condition with 33.2 kW of fuel cell peak power.

9. References

[1] Jato, "Internal Combustion Engines (ICE) counted for over 90% of global car sales in H1 2019," 2019. .
[2] Alternative Fuels Data Center, "How Do Fuel Cell Electric Vehicles Work Using Hydrogen?" https://afdc.energy.gov/vehicles/how-do-fuel-cell-electric-cars-work.
[3] J. G. Hayes and G. A. Goodarzi, *Electric Powertrain*, vol. 148. .
[4] C. Hochgraf, "Encyclopedia of Electrochemical Power Sources," *Encycl. Electrochem. Power Sources*, pp. 236–248, 2009, [Online]. Available: http://www.sciencedirect.com/science/article/pii/B9780444527455008637.
[5] L. Jeongwoo and D. J. Nelson, "Rotating inertia impact on propulsion and regenerative braking for electric motor driven vehicles," *2005 IEEE Veh. Power Propuls. Conf. VPPC*, vol. 2005, pp. 308–314, 2005, doi: 10.1109/VPPC.2005.1554575.
[6] A. Karki, B. P. Shrestha, D. Tuladhar, S. Basnet, S. Phuyal, and B. Baral, "Parameters Matching for Electric Vehicle Conversion," *2019 IEEE Transp. Electrif. Conf. ITEC-India 2019*, pp. 1–5, 2019, doi: 10.1109/ITEC-India48457.2019.ITECIndia2019-133.
[7] F. Barbir, "Fuel Cell Vehicle," *Hydrog. Energy Syst.*, pp. 241–251, 1995, doi: 10.1007/978-94-011-0111-0_16.
[8] D. Ramos Matos, "UC Irvine UC Irvine Electronic Theses and Dissertations Publication Date," 2018, [Online]. Available: https://escholarship.org/uc/item/1tr0c2p0.
[9] M. İnci, M. Büyük, M. H. Demir, and G. İlbey, "A review and research on fuel cell electric vehicles: Topologies, power electronic converters, energy management methods, technical

challenges, marketing and future aspects," *Renew. Sustain. Energy Rev.*, vol. 137, no. March 2020, 2021, doi: 10.1016/j.rser.2020.110648.

[10] D. Feroldi, M. Serra, and J. Riera, "Design and analysis of fuel-cell hybrid systems oriented to automotive applications," *IEEE Trans. Veh. Technol.*, vol. 58, no. 9, pp. 4720–4729, 2009, doi: 10.1109/TVT.2009.2027241.

[11] E. M. Allam and S. M. Metwalley, "Fuel cell electric vehicle simulation in Simulink and its use for power management studies," *Int. J. Electr. Hybrid Veh.*, vol. 4, no. 4, pp. 297–313, 2012, doi: 10.1504/IJEHV.2012.053023.

[12] B. Duangsrikaew, J. Mongkoltanatas, P. Karin, K. Hanamura, and C. Benyajati, "Battery sizing for electric vehicles based on real driving pattern in Thailand," *31st Int. Electr. Veh. Symp. Exhib. EVS 2018 Int. Electr. Veh. Technol. Conf. 2018, EVTeC 2018*, 2018.

[13] DieselNet, "Worldwide Harmonized Light Vehicles Test Cycle (WLTC)," *ECOpoint Inc.*, 2019. https://dieselnet.com/standards/cycles/wltp.php#:~:text=WLTC Test Cycles,-Class 3 Cycle&text=Class 3 vehicles are divided,v_max ≥ 120 km%2Fh. (accessed May 05, 2022).

[14] Cardekho, "Specifications of Maruti Alto 800," *Girnar Software Pvt. Ltd.* https://www.cardekho.com/maruti/maruti-alto-800-specifications.htm (accessed May 05, 2022).

[15] K. Sarrafan, D. Sutanto, K. M. Muttaqi, and G. Town, "Accurate range estimation for an electric vehicle including changing environmental conditions and traction system efficiency," *IET Electr. Syst. Transp.*, vol. 7, no. 2, pp. 117–124, 2017, doi: 10.1049/iet-est.2015.0052.

[16] S. Ramireddy, M. R. Dandu, and V. R. Prasad, "Fabrication and Performance Analysis of Prototype PEM Fuel Cell Hybrid Vehicle," *Int. J. Eng. Tech.*, vol. 4, no. 3, 2018.

[17] MAGNET-SCHULTZ, "High-pressure cut-off valve Type E-G 032-1110 G PC." https://www.magnet-schultz.com/en/valves-for-hydrogen-storage-and-fuel-cell-systems/high-pressure-cut-off-valve/.

[18] J. L. Hall, "Cell components," *Phytochemistry*, vol. 26, no. 4, pp. 1235–1236, 1987, doi: 10.1016/s0031-9422(00)82398-5.

[19] S. Chauhan, "Motor Torque Calculations For Electric Vehicle," *Int. J. Sci. Technol. Res.*, vol. 4, no. 8, pp. 126–127, 2015.

[20] S. Manish and R. Banerjee, "Techno-economic assessment of fuel cell vehicles for India," *16th World Hydrog. Energy Conf. 2006, WHEC 2006*, vol. 1, no. January 2006, pp. 322–332, 2006.

[21] J. Larminie and A. Dicks, "Appendix 2: Useful Fuel Cell Equations," *Fuel Cell Syst. Explain.*, pp. 395–400, 2013, doi: 10.1002/9781118878330.app2.

Acknowledgement
We are grateful for the insightful support of our supervisors at Green Hydrogen Lab, Kathmandu University.

Knee Flexion/Extension Angle Measurement for Gait Analysis Using Machine Learning Solution "MediaPipe Pose" and Its Comparison with Kinovea®

Amit Gupta[1], Pratisthit Lal Shrestha[1*], Bhola Thapa[1], Raman Silwal[1], Raju Shrestha[1]
[1]Design Lab, Department of Mechanical Engineering Kathmandu University, Nepal
*Corresponding author: pratisthit@ku.edu.np

Abstract. Human Gait Analysis is crucial in diagnosis, monitoring, treatment, and rehabilitation. It requires the measurement of gait parameters like joint angles, angular motion, ground reaction force (GRF), momentum, step width, step length, velocity, etc. This study introduces a markerless, easy-to-use approach for measuring one of the gait parameters i.e., knee flexion/extension angle, and verifies it with an existing standard marker-based approach. Knee flexion/extension angle is calculated via Machine Learning (ML) pose estimation model "MediaPipe Pose" and Computer vision (CV) without the use of markers. For the verification of accuracy, the obtained values are compared with the data on each video frame obtained from Kinovea® which is a marker-based motion analysis software. A correlation of 0.941 was observed between the results from Kinovea® and MediaPipe Pose. Similarly, the mean absolute error of knee angles was 5.88 degrees. The research shows that knee flexion/extension angle can be accurately measured using ML solution for high-fidelity body pose tracking. Professionals involved in the field of biomechanics, sports medicine, physiotherapy as well as other medical fields can use this method as an alternative markerless approach for knee flexion/extension angle measurement for gait analysis.

Keywords: Machine Learning, Gait Analysis, Knee flexion/extension angle, Pose Estimation, MediaPipe Pose, Kinovea®

1. Introduction

Human gait represents a person's manner of walking. Human movement is the result of a series of events repeated by each limb. A single sequence of events by one limb constitutes one gait cycle. The gait cycle means the duration between the consecutive occurrence of one of the recurring events of walking. Initial contact between the person's foot and the ground is taken as the reference recurring event. In a normal gait cycle, a heel strike is generally the initial contact [1].

Human gait is a significant indicator of health as it is helpful in a broad range of situations such as neurological diseases, diabetes, and fall detection and prediction. Its quantification can be done using gait analysis systems. Gait analysis can be referred to as the scientific study of the gait cycle and it plays an important role in diagnosing musculoskeletal and neurological disorders, as well as assessing the effectiveness of various interventions performed on patients [2].

Gait analysis and evaluation require the measurement of Kinematic parameters (such as joint angles, angular motion, etc.), Kinetic parameters (such as Ground Force Reaction (GRF), momentum, etc.), Spatio-Temporal parameters (such as velocity, step width, step length, etc.), Anthropometric parameters (such as weight, height, age) and much more [3]. Knee flexion/extension angle is one such gait parameter.

Various methods have been proposed for gait analysis ranging from 2D to 3D, marker-based to markerless, and vision-based to sensor-based [3]–[6]. Three-dimensional motion analysis methods are regarded as the gold standard for gait analysis as they produce quantitative and objective data concerning spatiotemporal and kinematic parameters [7]. However, motion analysis systems have some disadvantages like the need for trained manpower, high cost, high processing time, and considerable space for installation and setup. Marker-based gait analysis systems have noise in the evaluated gait parameters because the markers placed on the subject's body are susceptible to movements produced from the sliding of skin over the bones [8].

Gait parameters can also be evaluated from 2D video using various software. Various study shows that motion analysis software like Kinovea®, Templo, etc. are reliable for measuring the kinematic parameters [9]–[11]. Recent Development in the field of computer vision and machine learning have assisted in understanding digital images and videos. For the analysis and study of digital images, Convolutional Neural Networks (CNN) are designed and used [12]. CNN is preferred in a wide range of applications like edge detection, image classification, facial recognition, and human pose estimation [13].

In this study, a markerless, easy-to-use method for gait analysis is suggested. MediaPipe Pose, an ML solution for markerless 3D human pose estimation is used to measure knee flexion/extension angles. Video recorded from a mobile phone camera is analyzed using the 3D pose estimation system to get 33 important 3D anatomical landmarks. Using the hip, knee, and ankle landmarks, the knee flexion/extension angle is measured. The effectiveness of this method is tested by comparing the obtained knee angles with Kinovea® [14].

2. Materials And Methods

Human pose estimation[15] can be defined as a process of figuring or estimating human poses from a digital image. Pose estimation includes accurate detection and identification of human joints. Pose estimation algorithms approaches can be categorized into two types, top-down and bottom-up. The top-down approach [16] involves first finding possible human regions from the image and then extracting joints from those regions and finally inferring poses from the extracted joints. In the top-down approach, as each region is processed individually, it can be computationally very expensive. The bottom-up approach [17] involves processing the whole image to extract possible joints and then these joints are connected to generate a pose. This approach is computationally less expensive than the top-down approach. However, its drawback is that multi-person pose estimation can produce faulty pose models due to the erroneous connection of joints.

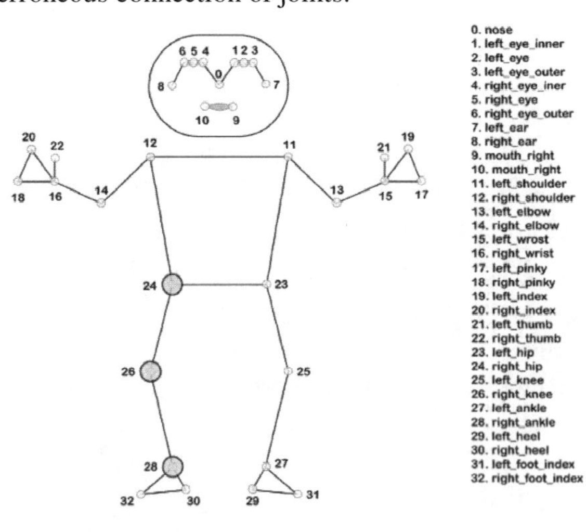

Figure 1. 33 Pose landmarks in MediaPipe Pose Landmark Model

IOP Conf. Series: Materials Science and Engineering 1279 (2023) 012004 doi:10.1088/1757-899X/1279/1/012004

MediaPipe is a framework or platform for constructing machine learning pipelines for handling and processing time-series data like audio, video, etc. MediaPipe offers various machine-learning solutions like face detection, object detection along with pose. MediaPipe Pose is one of the ML solutions in the MediaPipe framework. MediaPipe Pose is a bottom-up multi-person pose estimation model. It can detect 33 anatomical landmarks. Figure 1 shows that landmark numbers 24(Right Hip), 26(Right Knee), and 28(Right Ankle) are used to measure the right leg's knee flexion/extension angle during a gait cycle. Kinovea® is open-access 2D video analysis software, available online (https://www.kinovea.org/). It can be used as a motion analysis tool [18], [19] as well as an accurate method for gait analysis [8]. Kinovea® software helps to analyze video without markers, however, the use of passive markers might improve the reliability [20].

In this experiment, video recordings of a walking subject with markers on the hip, knee, and ankle of the right leg were captured. However, pose estimation does not require markers, but they are used to analyze the videos with Kinovea®. Gait cycles were captured using a mobile phone camera mounted on a tripod stand. The phone resolution was 1380 x 1200 pixels at an average rate of 30 frames per second. A single healthy subject aged 24 participated in the experiment. Eight (8) videos consisting of single gait cycles of the subject were recorded for the experiment. The camera height was 1000 mm and the distance between the camera and the subject was 2000 mm. Then the videos were analyzed by MediaPipe Pose and Kinovea®. For each frame during the gait cycle for each trial, the knee angles were calculated using both methods. Hip, knee, and ankle coordinates were obtained from the pose estimation MediaPipe Pose model using a Python program to calculate the knee angles. Similarly, the knee angle was also calculated using the Kinovea® software for comparison. RMS (Root Mean Square) values were then calculated for the obtained knee angles. Root Mean Square (RMS) is the statistical evaluation of the magnitude of diverse quantities. Then the RMS values from both these methods were used to calculate the Pearson correlation coefficient (r) and Mean Absolute Error (MAE). Pearson correlation is the measure of how strongly two variables are related. Its value is a number between -1 and +1, with +1 representing a positive correlation. Mean Absolute Error (MAE) is the evaluation of errors between paired studies demonstrating an identical phenomenon. Figure 2 and Figure 3 show the knee angle calculated for the same instance using both methods and described approach.

Figure 2. Knee angle evaluated by MediaPipe Pose

Figure 3. Knee angle evaluated by Kinovea®

3. Results and Discussions

The average correlation coefficient and mean absolute error of the experiment were calculated. High correlations (r = 0.941) between Kinovea® and MediaPipe Pose were observed. Similarly, the mean absolute error of knee angles between MediaPipe Pose and Kinovea® was 5.88 degrees.

Table 1: Pearson Correlation Coefficient for each 8 Trials

	Trial1	Trial2	Trial3	Trial4	Trial5	Trial6	Trial7	Trial8
Correlation Coefficient(r)	0.953	0.940	0.939	0.963	0.939	0.944	0.923	0.924

Table 2: Mean Absolute Error for each 8 Trials

	Trial1	Trial2	Trial3	Trial4	Trial5	Trial6	Trial7	Trial8
Mean Absolute Error (Degree)	5.586	4.549	5.929	4.769	5.036	5.093	5.900	5.439

It is observed that mean absolute error is most during the "Loading Response" and "Mid-Swing" phases of the gait cycle. The maximum knee flexion/extension angle measured using Kinovea® is always less than MediaPipe Pose for those phases. A possible reason for this might be the slightly imprecise placement of markers on clothes/pants rather than on skin, which might have produced imprecise angles using Kinovea®, especially during those phases of the gait cycle. Overall, the high correlation and low mean absolute error suggest that pose estimation could be a correct approach and MediaPipe Pose could be a valid tool for measuring knee flexion/extension angles for gait analysis.

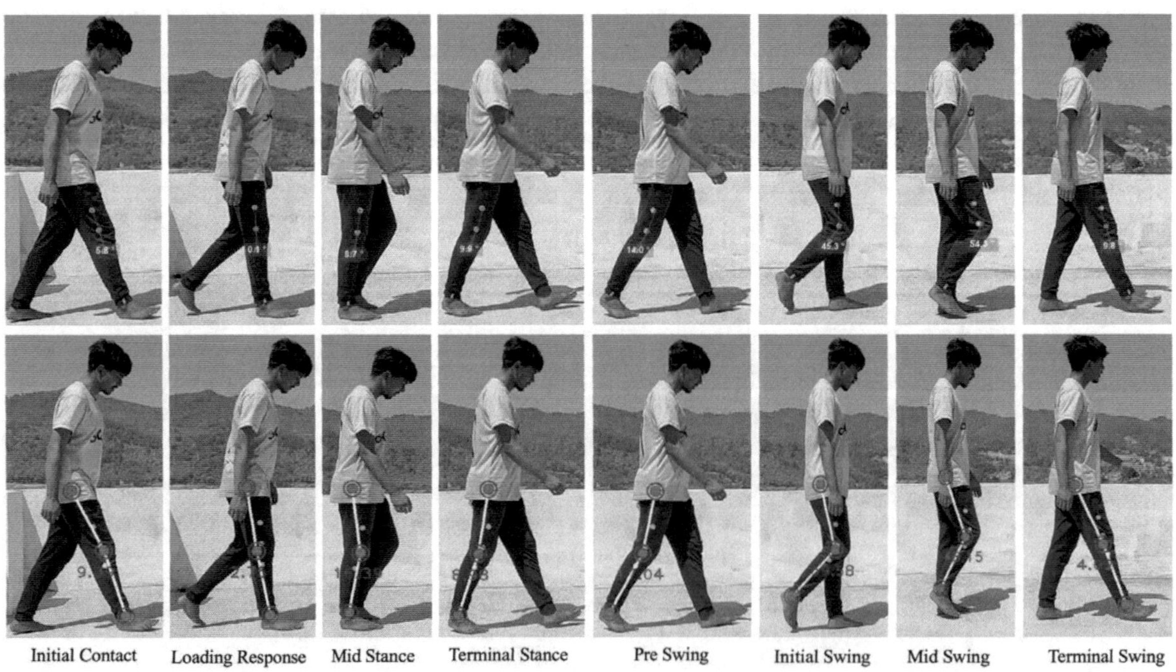

Figure 4. Knee flexion/extension angle during the 8 phases of the gait cycle

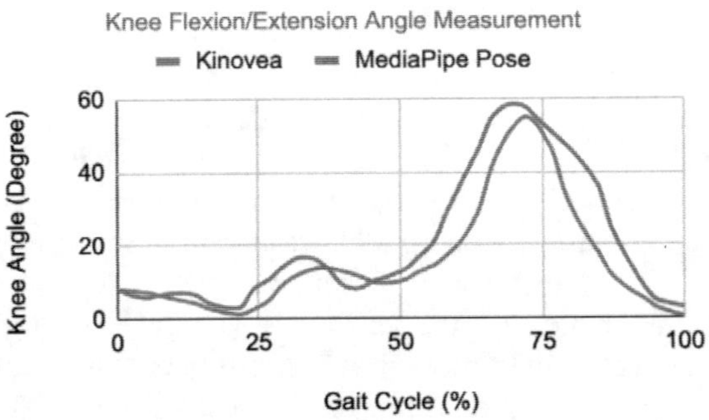

Figure 5. Knee angles for Trial 1

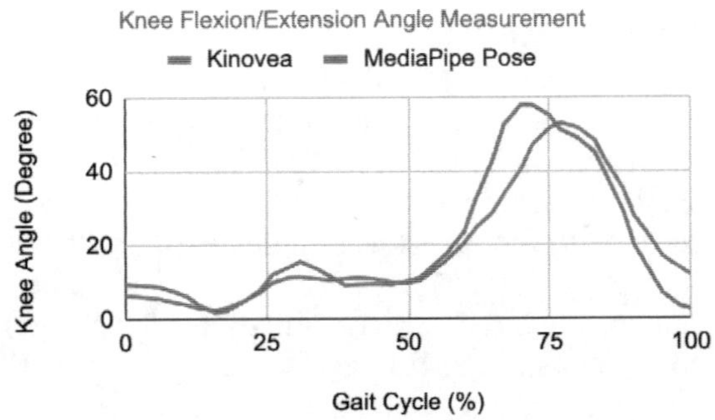

Figure 6. Knee angles for Trial 2

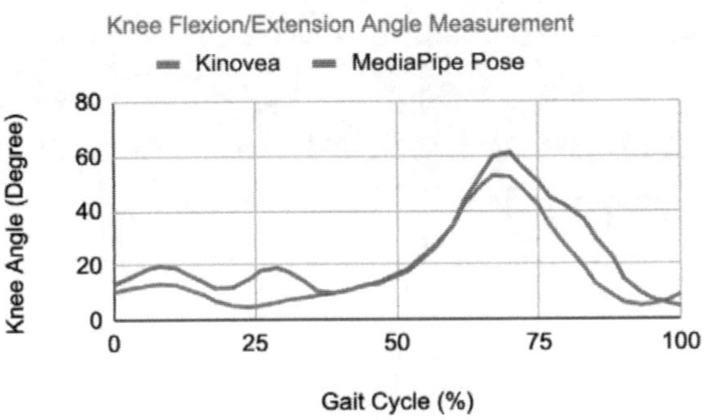

Figure 7. Knee angles for Trial 3

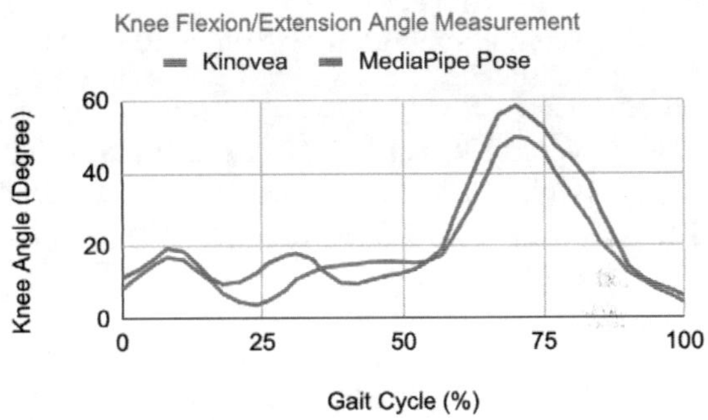

Figure 8. Knee angles for Trial 4

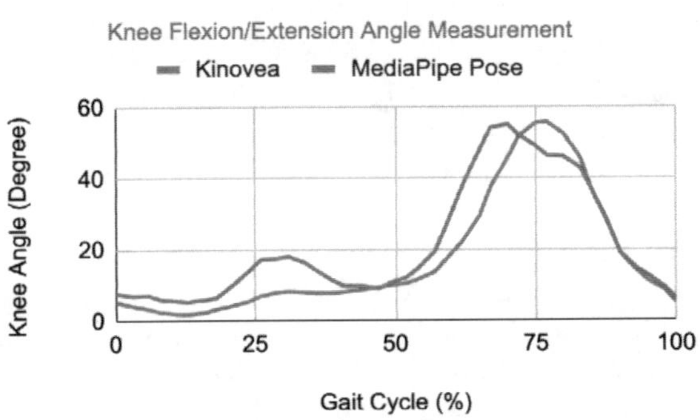

Figure 9. Knee angles for Trial 5

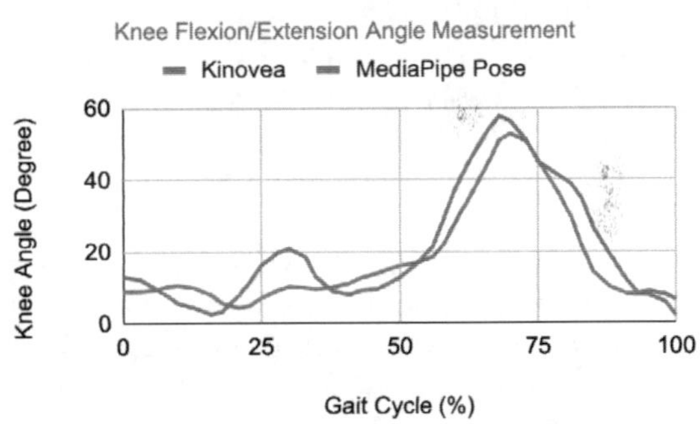

Figure 10. Knee angles for Trial 6

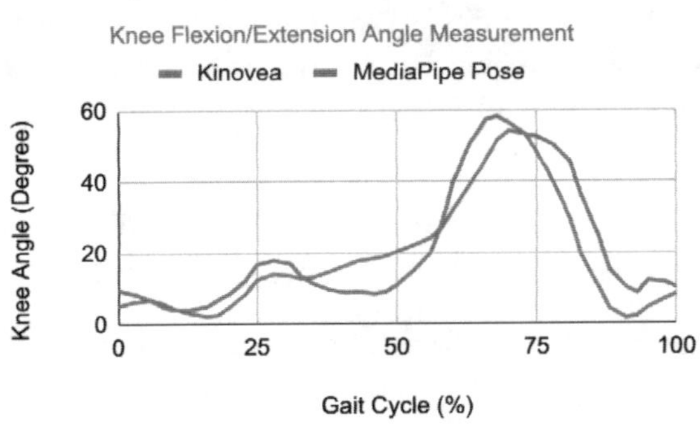

Figure 11. Knee angles for Trial 7

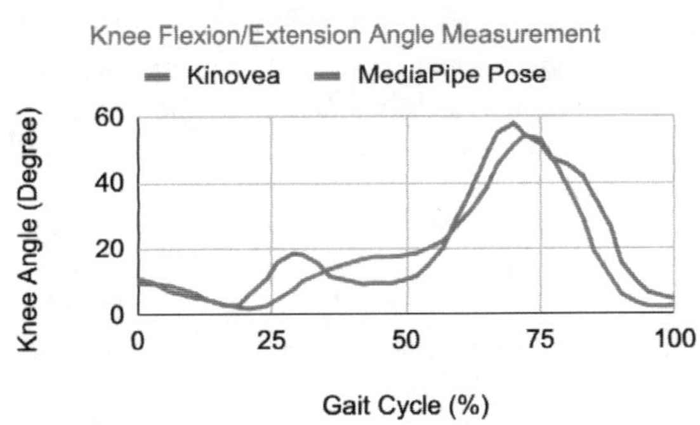

Figure 12. Knee angles for Trial 8

4. Conclusions

High correlation (r = 0.941) and low mean absolute error (5.88 degrees) between the two methods show that gait parameters like knee flexion/extension angle can be accurately measured using pose estimation techniques like "MediaPipe Pose". It also shows that 2D video gait analysis is possible by obtaining anatomical landmarks using the pose estimation model. The pose estimation approach can be used to measure knee angles easily without the use of any markers. This research shows that a simple phone camera and a personal computer can be used to measure a gait parameter. It is also demonstrated that when comparing MediaPipe Pose with Kinovea® the results are accurate. Though pose estimation could have some minor limitations related to lighting and the environment in which the experiment is conducted, overall, the research provides a markerless, easy-to-use approach for measuring gait parameters namely knee flexion/extension angle. Also, based on this research other kinematic gait parameters like hip angles, ankle angles, stride length, velocity, etc. could be measured.

ACKNOWLEDGEMENT

This research is supported by Simula Research Laboratory, Norway.

REFERENCES

[1] A. Kharb, V. Saini, Y. K. Jain, and S. Dhiman, "A review of gait cycle and its parameters," 2011. [Online]. Available: www.IJCEM.orgIJCEMwww.ijcem.org
[2] A. Harvey and J. W. Gorter, "Video gait analysis for ambulatory children with cerebral palsy: Why, when, where and how!," Gait Posture, vol. 33, no. 3, pp. 501–503, Mar. 2011, doi: 10.1016/j.gaitpost.2010.11.025.
[3] C. Prakash, R. Kumar, and N. Mittal, "Recent developments in human gait research: parameters, approaches, applications, machine learning techniques, datasets and challenges," Artif Intell Rev, vol. 49, no. 1, Jan. 2018, doi: 10.1007/s10462-016-9514-6.
[4] Y. Matsushita, D. T. Tran, H. Yamazoe, and J.-H. Lee, "Recent use of deep learning techniques in clinical applications based on gait: A survey," J Comput Des Eng, vol. 8, no. 6, pp. 1499–1532, 2021.
[5] L. Lonini et al., "Video-based pose estimation for gait analysis in stroke survivors during clinical assessments: a proof-of-concept study," Digit Biomark, vol. 6, no. 1, pp. 9–18, 2022.
[6] V. Cimolin et al., "Computation of gait parameters in post stroke and parkinson's disease: A comparative study using RGB-D sensors and optoelectronic systems," Sensors, vol. 22, no. 3, p. 824, 2022.
[7] J. J. Daly et al., "Development and testing of the Gait Assessment and Intervention Tool (G.A.I.T.): A measure of coordinated gait components," J Neurosci Methods, vol. 178, no. 2, pp. 334–339, Apr. 2009, doi: 10.1016/j.jneumeth.2008.12.016.
[8] C. Reinschmidt, A. J. van den Bogert, B. M. Nigg, A. Lundberg, and N. Murphy, "Effect of skin movement on the analysis of skeletal knee joint motion during running," J Biomech, vol. 30, no. 7, pp. 729–732, 1997.
[9] P. Fernández-González, A. Koutsou, A. Cuesta-Gómez, M. Carratalá-Tejada, J. C. Miangolarra-Page, and F. Molina-Rueda, "Reliability of kinovea® software and agreement with a three-dimensional motion system for gait analysis in healthy subjects," Sensors (Switzerland), vol. 20, no. 11, Jun. 2020, doi: 10.3390/s20113154.
[10] N. A. H. Hisham, A. F. A. Nazri, J. Madete, L. Herawati, and J. Mahmud, "Measuring ankle angle and analysis of walking gait using kinovea," 2017.
[11] M. Hamdan, S. I. Ismail, H. Hassan, H. Ismail, S. A. Bukry, and R. M. F. Raja Azidin, "How reliable is kinovea vs. templo in knee and hip kinematics assessment during side cutting tasks," Knee, vol. 2, no. 1, 2017.
[12] W. Rawat and Z. Wang, "Deep convolutional neural networks for image classification: A comprehensive review," Neural Computation, vol. 29, no. 9. MIT Press Journals, pp. 2352–2449, Sep. 01, 2017. doi: 10.1162/NECO_a_00990.
[13] M. Egmont-Petersen, D. de Ridder, and H. Handels, "Image processing with neural networks-a review," 2002. [Online]. Available: www.elsevier.com/locate/patcog

[14] A. Puig-Diví, J. M. Padullés-Riu, A. Busquets-Faciaben, X. Padullés-Chando, ; Carles Escalona-Marfil, and D. Marcos-Ruiz, "Validity and Reliability of the Kinovea Program in Obtaining Angular and Distance Dimensions," 2017, doi: 10.20944/preprints201710.0042.v1.
[15] R. Josyula and S. Ostadabbas, "A Review on Human Pose Estimation," Oct. 2021, [Online]. Available: http://arxiv.org/abs/2110.06877
[16] G. Ning, P. Liu, X. Fan, and C. Zhang, "A top-down approach to articulated human pose estimation and tracking," in Proceedings of the European Conference on Computer Vision (ECCV) Workshops, 2018, p. 0.
[17] M. Li, Z. Zhou, J. Li, and X. Liu, "Bottom-up pose estimation of multiple person with bounding box constraint," in 2018 24th International Conference on Pattern Recognition (ICPR), 2018, pp. 115–120.
[18] S. H. Elwardany, W. H. El-Sayed, and M. F. Ali, "Reliability of Kinovea Computer Program in Measuring Cervical Range of Motion in Sagittal Plane," OAlib, vol. 02, no. 09, pp. 1–10, 2015, doi: 10.4236/oalib.1101916.
[19] C. H. Guzmán-Valdivia, A. Blanco-Ortega, M. A. y Oliver-Salazar, and J. L. Carrera-Escobedo, "Therapeutic motion analysis of lower limbs using Kinovea," Int J Soft Comput Eng, vol. 3, no. 2, pp. 2231–2307, 2013.
[20] C. Damsted, R. O. Nielsen, and L. H. Larsen, "Reliability of video-based quantification of the knee-and hip angle at foot strike during running," Int J Sports Phys Ther, vol. 10, no. 2, p. 147, 2015.

Parametric Modelling and Optimization of Alkaline Electrolyzer for the Production of Green Hydrogen

S Niroula[1], C Chaudhary[2], A Subedi[1,3] and B S Thapa[1,3*]

[1] Department of Mechanical Engineering, Kathmandu University, Nepal
[2] Department of Electrical and Electronics Engineering, Kathmandu University, Nepal
[3] Green Hydrogen Lab, Kathmandu University, Nepal
* Corresponding Author: bst@ku.edu.np

Abstract. Electrolysis is the process of decomposition of the liquid, containing ions by passing electricity through it. This process is achieved by an electrolyzer. An alkaline water electrolyzer uses electrical energy to break the chemical bond between hydrogen and oxygen in an alkaline electrolytic medium. The performance of the electrolyzer depends upon various parameters like temperature, pressure, ohmic overpotential, activation overpotential, etc. In this study, various parameters that affect the performance of the electrolyzer are identified and integrated using fundamental thermodynamics and electrochemical equations. A mathematical model is created in MATLAB/SIMULINK to analyse the voltage vs current density plot of the electrolyzer. An attempt is made to optimize the performance of the electrolyzer through the variation of the parameters. It was observed that increasing the operating temperature and decreasing the operating pressure favours the electrolysis process. Also, 30 wt% of KOH solution as electrolyte exhibits the optimal cell voltage as compared to 20 and 40 wt% solutions.

Keywords: Green Hydrogen, Electrolysis, Alkaline Electrolyzer, Parametric Modelling.

1. Introduction

Hydrogen is the lightest and the most abundant element in the universe. It has been in use for a long for manufacturing ammonia and methanol and also for refining petroleum products[1]. But with the increasing concern related to global warming and other environmental issues, hydrogen is emerging as an energy vector to channel the energy from renewable sources, and meet the continuous demand of large-scale consumption in recent two decades. Hydrogen can be used with fuel cell technology to produce energy with no carbon emission and minimal emission of other pollutants. It has the advantage of having the highest energy density among other fuels i.e. 120 MJ/kg which is more than twice the energy density of other fossil fuels [2]. There are numerous ways of producing hydrogen, like hydrocarbon reforming, hydrocarbon pyrolysis, electrolysis etc.[3]. Currently, 96% of the hydrogen produced worldwide, is based on fossil fuels[4]. The state-of-art commercialized alternative for producing green hydrogen using renewable sources is electrolysis of water which uses electricity produced from renewable sources like solar, wind, hydropower etc.

The electrolysis of water is the splitting of water (H_2O) into hydrogen (H_2) and oxygen (O_2) gas with the application of electricity energy. The basic chemical reaction of electrolysis of water is shown in equation (1).

$$H_2O + energy \rightarrow H_2 + \frac{1}{2}O_2 \tag{1}$$

The device that performs electrolysis is called an electrolyzer. The electrolyzer is around 70% efficient, produces highly pure hydrogen, has few moving parts, and needs minimal space[5]. There are mainly three types of electrolyzer: proton exchange membrane (PEM), alkaline, and solid oxide electrolyzer (SOE). Alkaline and PEM electrolyzer are already commercialized whereas solid oxide electrolyzer are still in the research phase. Alkaline electrolysis is the most widely utilized process for producing hydrogen from water because of its high maturity and low costs compared to other methods of electrolysis[6]. In an alkaline electrolyzer, water is split into hydrogen and oxygen in an alkaline electrolytic medium, commonly 20-30 wt% KOH solution. The schematic of an alkaline electrolyzer is shown in figure 1. A cell frame, electrolyte, anode and cathode electrode, and separating diaphragm are the major components of an alkaline electrolyzer. It can operate in pressure ranging from 1-30 bars and has an operating temperature of 60-80°c [6]. Each cell of an alkaline electrolyzer requires a minimum of 1.229 V at STP[7]. In this paper parametric modelling of alkaline electrolyzer is discussed. The result polarization curve is generated on the basis of model and validated against already published data. The overall chemical reaction taking place in the anode and cathode side of an alkaline electrolyzer is presented in equations (2) and (3) respectively.

$$At\ anode: \quad 2OH^-(aq) \rightarrow \frac{1}{2}O_2(g) + H_2O(l) + 2e^- \tag{2}$$

$$At\ cathode: \quad 2H_2O(l) + 2e^- \rightarrow H_2(g) + 2OH^-(aq) \tag{3}$$

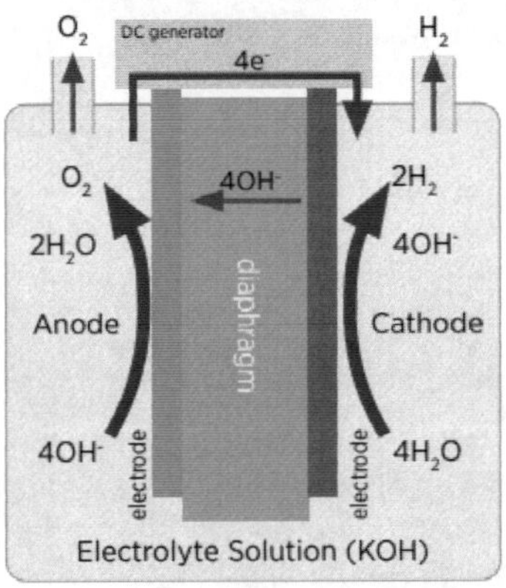

Figure 1. Schematic of an Alkaline Electrolyzer Cell [6]

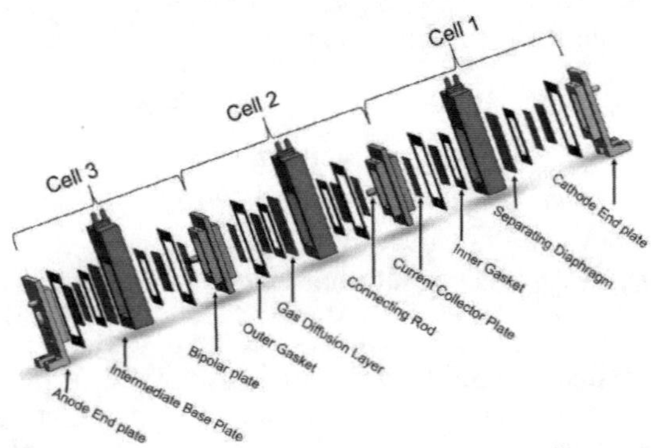

Figure 2. Three Cell Alkaline Electrolyzer Stack [8]

A single cell of an alkaline electrolyzer can operate at 1.8 - 2.4 volts depending upon operating condition[9]. To increase to operating voltage, single cell is connected in series as shown in Figure 2 to form of a stack of electrolyzer. The performance of a electrolyzer is characterized and compared on the basis of a polarization curve. A Polarization curve is the curve plotted with current density on x-axis and cell voltage in y-axis.

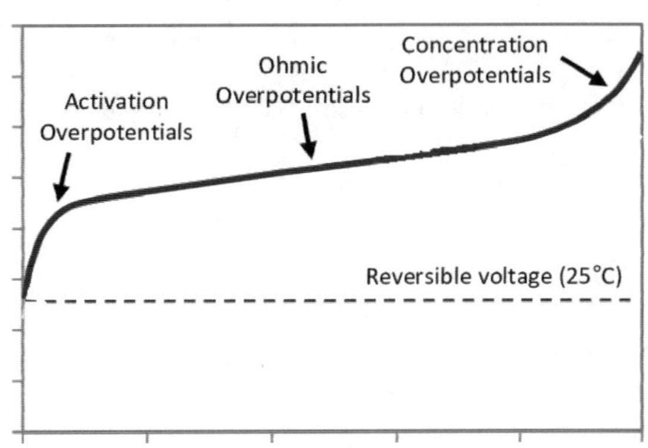

Figure 3. A Typical Polarization Curve of an Alkaline Electrolyzer Cell [10]

IOP Conf. Series: Materials Science and Engineering 1279 (2023) 012005 doi:10.1088/1757-899X/1279/1/012005

A typical polarization curve of an alkaline electrolyzer is shown in Figure 3. It has 3 distinct regions i.e., activation region at lower current density, ohmic region at moderate current density and concentration region at high current density. At lower current density the sudden rise in voltage due to overpotential caused by activation of electrode whereas in in moderate current density ohmic overpotential are dominant. At higher current density voltage again steeply increased due to concentration overpotential.

2. Mathematical Modelling

A Mathematical model based on various parameters are used to generate a Polarization curve of an electrolyzer. Thus, formed polarization curve is validated against experimental data. To develop the mathematical model, different thermodynamical, electrochemical, and empirical relations are used. Different parameters that affect the performance of the electrolyzer are linked together with those equations and finally output cell voltage is obtained.

2.1. Cell voltage

The cell voltage of an alkaline electrolyzer cell is the actual voltage that an electrolyzer operates on. It considers all of the losses inside the electrolyzer cell that occurs during the operation of the electrolyzer. It can be determined using equation (4) which is sum of reversible voltage and all the overpotential occurring in electrolyzer [11].

$$V_e = V_{rev} + V_{act} + V_{ohmic} + V_{conc} \tag{4}$$

- V_{rev} is reversible voltage
- V_{act} is activation overvoltage
- V_{ohmic} is ohmic overvoltage
- V_{conc} is concentration overvoltage

Concentration losses occurs at very high current density (i.e., greater than 800 mA/cm^2) which is much higher than operating range of current density of an alkaline electrolyzer. Hence many authors [12], [13], [14] have neglected the effect of concentration or mass transfer overpotential in modelling of the electrolyzer. Concentration overpotential is neglected also in this study.

2.2. Reversible voltage

The reversible voltage or open circuit voltage is ideal voltage required start the electrolysis process. It depends upon operating temperature and pressure of the electrolysis process and is determined using the Nernst equation. The modified Nernst equation with an additional term to account approximately for the change in the reversible voltage due to change of temperatures is given by equation (5) [15][16].

$$V_{rev} = E^{\circ}{}_{rev} + (T - T_{ref}) * \frac{\Delta S^{\circ}}{nF} + \frac{RT}{2F}\left[\ln\left(\frac{(P - P_{H_2O})^{\frac{3}{2}}}{\frac{P_{H_2O}}{P^{\circ}{}_{H_2O}}}\right)\right] \tag{5}$$

- $E^{\circ}{}_{rev}$ is the minimum reversible voltage required for electrolysis reaction to occur at STP.
- T is operating temperature.
- T_{ref} is the reference temperature at which $E^{\circ}{}_{rev}$ and ΔS° is evaluated.
- $\Delta S^{\circ}/nF$ is the standard state entropy change at 25° C.
- R is universal gas constant.
- F is faraday constant.
- P is the operational pressure.
- $P_{H_2O}$is vapor pressure of water over the electrolyte.
- $P^{\circ}{}_{H_2O}$ vapor pressure of pure H_2O.

The minimum reversible voltage required for electrolysis reaction to occur at STP is given by equation (6)

$$E^{\circ}_{rev} = \frac{\Delta G^{\circ}}{2F} \tag{6}$$

- ΔG° is Gibbs free energy change, and is given by equation (7).

$$\Delta G^{\circ} = \Delta H^{\circ} - T_{ref}\Delta S \tag{7}$$

- ΔH° is the change in enthalpy at T_{ref}.
- ΔS° is the change in entropy at T_{ref}.

The vapor pressure of water over the electrolyte can be calculated using (8) [17] and vapor pressure of pure H_2O is given by equation (9).

$$\ln P_{H_2O} = 0.016214 - 0.13802m + 0.19330\sqrt{m} + 1.0239 \ln P^{\circ}_{H_2O} \tag{8}$$

$$\ln P^{\circ}_{H_2O} = 37.043 - \frac{6275.7}{T} - 3.4159 \ln T \tag{9}$$

Where,
m is molarity of the electrolyte and can be calculated using equation (10)[18][19].

$$m = \frac{Wt\%\left(183.1221 - 0.56845T + 984.5679 \exp\left(\frac{wt\%}{115.96277}\right)\right)}{5610.5} \tag{10}$$

- wt. % is weight percentage of KOH solution.

2.3. Activation overpotential

The activation overpotential is a measure of the energy required to break chemical bonds prior to an electrochemical reaction. It results of the kinetics of the electronic charge transfer reactions that take place at the reaction site of electrode[20]. Activation overpotential is the function of current density and exchange current density. Current density is the current of electron or ions per unit area of the surface. The exchange current density is the current in the absence of net electrolysis and at zero overpotential. The activation voltage is the result of an overvoltage across the electrode [11]. Activation overpotential can be determined using Tafel equation [21] and is given by equations (11) and (12) for anode activation and cathode respectively. The slope of activation overpotential in linear region of polarization curve is determined by Tafel slope. During the evolution of gas bubbles are also formed which get accumulated on the surface of electrode before they get detached. The bubble covering decreases the active area of the electrodes, lowering the exchange current density and increasing the activation voltage [22]. Hence, the effect of bubble coverage factor must also be considered while calculating the overpotential of the cell [23][24].

$$V_{act-a} = b_a \ln\left(\frac{j_a}{j_{o-a}}\right) \tag{11}$$

$$V_{act-c} = b_c \ln\left(\frac{j_c}{j_{o-c}}\right) \tag{12}$$

- $b_{a,c}$ is Tafel slope
- $j_{a,c}$ is the current density of electrodes
- $j_{o-a,c}$ is effective exchange current density.

Tafel slope for anode and cathode electrode is given by equation (13) [25].

$$b_{a,c} = \frac{RT}{nF\alpha_{a,c}} \tag{13}$$

- n is number of electrons
- $\alpha_{a,c}$ is the charge transfer coefficient of anode and cathode.

The charge transfer coefficient ($\alpha_{a,c}$) is determined by fitting the experimentally obtained data. In this study, the correlation mentioned by [19] for Ni electrode is used and is given by equation (14)(15).

$$\alpha_a = 0.0675 + 0.00095T \tag{14}$$

$$\alpha_c = 0.1175 + 0.00095T \tag{15}$$

The exchange current density $j_{o-a,c}$ is given by equation (16)[26].

$$j_{o-a,c} = \gamma_{a,c} * j_{o,ref-a,c} * \exp\left[\frac{-\Delta G_{c-a,c}}{R}\left(\frac{1}{T} - \frac{1}{T_{ref}}\right)\right] \tag{16}$$

- $\gamma_{a,c}$ is electrode roughness factor of electrode.
- $j_{o,ref-a,c}$ is reference exchange current density at temperature T_{ref}.
- $\Delta G_{c-a,c}$ is the free energy of activation.

The fractional bubble coverage (θ) on the electrode surface can be determined by using equation (17)[19][24][26].

$$\theta = \left(-97.25 + 182\frac{T}{T_{ref}} - 84\left(\frac{T}{T_{ref}}\right)^2\right)\left(\frac{j}{j_{lim}}\right)^{0.3}\frac{P}{P - P_{H_2O}} \tag{17}$$

- j_{lim} is the limiting current density at 100% bubble coverage[24].

Hence the activation overpotential given by equation (11) and (12) incorporating bubble coverage can be written as equation (18).

$$V_{act-a,c} = b_{a,c}\ln\left(\frac{j_{a,c}}{j_{o-a,c}}\right) + b_{a,c}\ln\left(\frac{1}{1-\theta}\right) \tag{18}$$

2.4. *Ohmic overpotential*

Ohmic overpotential is caused by electrical resistance in electrodes, electrolytes, and membranes. The total ohmic resistance is the sum of individual resistance caused by the electrode, electrolyte, and membrane, and the product of total ohmic resistance with current gives ohmic overpotential which is expressed as equation (19).

$$V_{ohmic} = I(R_{ele} + R_{ely} + R_{mem}) \tag{19}$$

- I is current in Ampere
- R_{ele} is resistance due electrode.
- R_{ely} is resistance due to electrolyte
- R_{mem} is resistance due to membrane.

2.4.1. Electrode resistance. Electrode resistance is the sum of electrical resistance due to anode and cathode electrode. The resistivity of the electrode depends upon the length, area, and resistivity of electrode material and is given by equation (20) [27].

$$R_{a,c} = \rho_{eff-a,c}\frac{t_{a,c}}{A_{a,c}} \tag{20}$$

- ρ_{eff} is the effective resistance of the electrode
- $t_{a,c}$ is the thickness of the electrode.
- $A_{a,c}$ is the cross-sectional area of the electrode.

The effective resistance of the electrode can be calculated using equation (21) [26].

$$\rho_{eff-a,c} = \frac{\rho_{o-a,c}}{(1-\epsilon_{e-a,c})^{\frac{3}{2}}} \tag{21}$$

- $\rho_{o-a,c}$ is the resistivity of the 100%-dense anode, cathode material at T_{ref}.
- ϵ_e is the porosity of the electrode

IOP Conf. Series: Materials Science and Engineering 1279 (2023) 012005 doi:10.1088/1757-899X/1279/1/012005

To compensate the change in effective resistance of 100%-dense anode, cathode material, due to the change in temperature, a new term k_{a-c} as temperature coefficient of resistivity is introduced. Hence, Then equation (20) can be written as equation (22) [26].

$$R_{ele} = \rho_{eff-a}\frac{t_a}{A_a}\left[1 + k_a(T - T_{ref})\right] + \rho_{eff-c}\frac{t_c}{A_c}\left[1 + k_c(T - T_{ref})\right] \tag{22}$$

2.4.2. Electrolyte resistance. Electrolyte resistance is the resistance possessed by electrolytic solution. The bubble formed also affect the resistance of the electrolyte solution. The total resistance electrolyte can be written as equation (23). It is the sum of electrolytic resistance with and without bubble [21].

$$R_{ely} = R_{ely-free} + R_{ely-bubble} \tag{23}$$

- $R_{ely-free}$ is resistance of bubble free electrolyte solution.
- $R_{ely-bubble}$ is resistance of electrolyte solution with bubble in it.

The electrolytic resistance with bubble is the function of bubble free electrolytic resistance and is given by equation (24)[21].

$$R_{ely-bubble} = R_{ely-free}\left(\frac{1}{\left(1 - \frac{2}{3}\theta\right)^{\frac{3}{2}}} - 1\right) \tag{24}$$

The bubble free electrolytic resistance is given by equation (25)

$$R_{ely-free} = \frac{1}{\sigma_{KOH,free}}\left(\frac{d_{am}}{A_a} + \frac{d_{cm}}{A_c}\right) \tag{25}$$

- $\sigma_{KOH,free}$ is bubble-free ionic conductivity of KOH solution.
- d_{am} is the distance between anode and membrane.
- d_{cm} is the distance between cathode and membrane.

The ionic conductivity of KOH can be determined by using empirical relation as equation (26) [28][29].

$$\sigma_{KOH-free} = -2.04m - 0.0028m^2 + 0.005332mT + 207.2\frac{m}{T} + 0.001043m^3 - 0.0000003\, m^2T^2 \tag{26}$$

2.4.3. Membrane resistance. A membrane or separator lies in the middle of two electrodes which maintain a continuous ionic conduction channel while preventing the mixing of gas. It contains several pores in it and the volume fraction of these pores is the porosity of the membrane (ϵ_m) [26]. The resistance due to membrane in conduction of ions depends upon various physical characteristics of the membrane and is given by equation (27) [26][28].

$$R_m = \left(\frac{1}{\sigma_{KOH,free}}\right)\frac{\tau_m^2 t_m}{\omega_m \epsilon_m A_m} \tag{27}$$

- τ_m is tortuosity of membrane.
- t_m is membrane thickness.
- ω_m is the wettability of membrane.
- ϵ_m is the porosity of the membrane.
- A_m is an area of membrane.

The tortuosity of porous materials refers to the lengthening of flow routes. It is the ratio between effective path lengths (the length of a single path from beginning to finish) and electrode thickness [30]. The wettability is the measure of the ability of a liquid to wet a surface of membrane [31].

3. Results and Discussion

The mathematical model using equations (4) – (27) is developed using MATLAB/SIMULINK. The constant parameters required for this model are presented in Table 1. The constants are based on the alkaline electrolyzer of Hydrogen Research Institute [21][26] [32].

Table 1. Constant Parameters based on HRI Electrolyzer [21][26] [32].

Constants	Symbol	Values	Unit
Reference Temperature	t_{ref}	298	K
Standard Entropy Change	$\Delta S°$	-162.86	J/mol.K
Number of electrons	n	2	
Faraday Constant	F	96485	C/mol
Universal Gas Constant	R	8.315	J/mol.K
Standard Enthalpy Change	$\Delta H°$	-286.02	Kj/mol
Weight Percent of KOH Solution	wt %	30	%
Electrode Roughness Factor Anode	γ_a	1.25	
Electrode Roughness Factor Cathode	γ_c	1.05	
Free Energy of Activation Anode	ΔG_a	41500	J/mol.K
Free Energy of Activation Cathode	ΔG_c	23450	J/mol.K
Reference Exchange Current Density Anode	$i_{o-ref,a}$	$1.34535 * 10^{-5}$	A/cm^2
Reference Exchange Current Density Cathode	$i_{o-ref,c}$	$1.8456 * 10^{-3}$	A/cm^2
Surface Area of Anode and Cathode	A_a, A_c	300	cm^2
Surface Area of Membrane	A_m	300	cm^2
Thickness of Anode and Cathode Electrode	t_a, t_c	0.2	cm
Thickness of Membrane	t_m	0.05	cm
Resistivity of the 100%-Dense NI Electrode	ρ_a, ρ_c	$6.4 * 10^{-6}$	Ω cm
Temperature Coefficient of Resistivity	k_a, k_c	0.00586	/°C
Distance Between Anode and Membrane	d_{am}	0.125	cm
Distance Between Cathode and Membrane	d_{cm}	0.125	cm
Porosity of Anode	ϵ_a	0.3	
Porosity of Cathode	ϵ_c	0.3	
Porosity of Membrane	ϵ_m	0.42	
Tortuosity of Membrane	τ_m	2.18	
Wettability Factor	ω_m	0.85	
Limiting current Density at 100 % bubble coverage	j_{lin}	30	A/cm^2

The voltage vs current density plot of a single cell of an alkaline electrolyzer based on the equation (4)-(25) and parameters mentioned in table 1 is plotted. The trend of the plot is compared to the experimental data of HRI electrolyzer from [21]. The experimental data points at different temperature and the corresponding polarization curve produced by the mathematical model can be found in Figure 4. The simulated result is observed to have good fit with the experimental data.

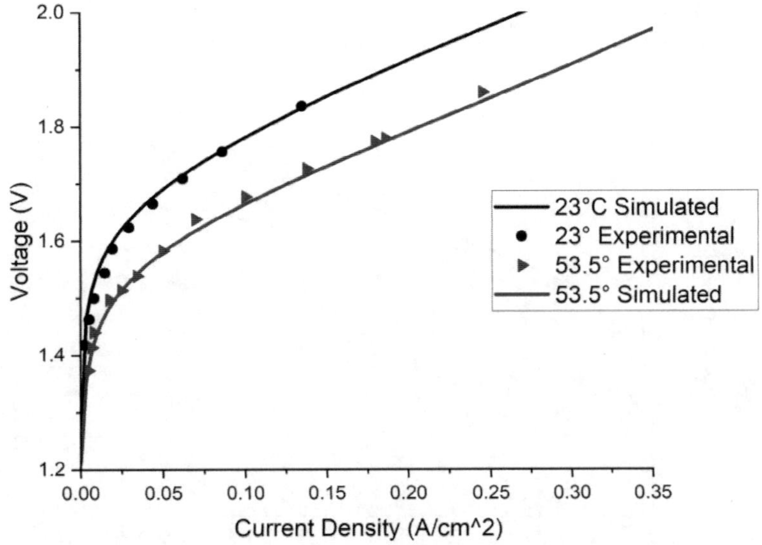

Figure 4. Comparison of simulated and experimental data of HRI electrolyzer at 1 bar pressures

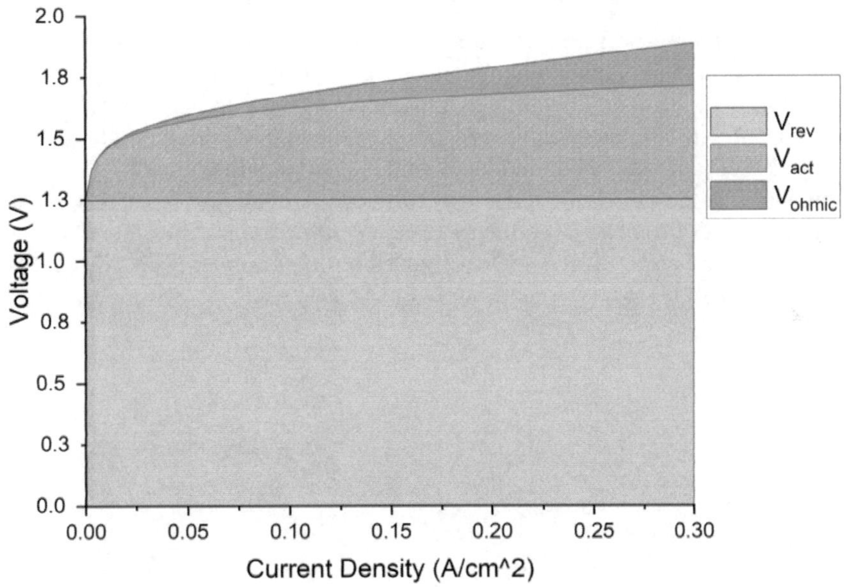

Figure 5. Cell voltage and Overpotential at 7 bar and 60° C

The total cell voltage with reversible voltage, activation overpotential, and ohmic overpotential vs current density plot is presented in Figure 5. It can be observed that the reversible voltage is constant through the polarization curve. It is only function of temperature and pressure and concentration of electrolyte as described in equation (5). For a given temperature, pressure and concentration of electrolyte, reversible voltage is a constant value. At low current density of polarization curve, activation overpotential rapidly increases which leads to steep rise in the overall cell voltage. Mathematically this rise is due to increase in rapid increase due logarithmic of ratio between current density and exchange current density at electrode. As ohmic overpotential is product of resistance and current, ohmic overpotential is very small and can be neglected in initial phase but as current density increases its effect becomes significant and shifts the polarization curves towards higher voltage value in later part of the graph.

3.1. *Effect of Temperature*

The variation of V-I curve with change in temperature at constant pressure of 7 bar is shown in **Figure 6**.

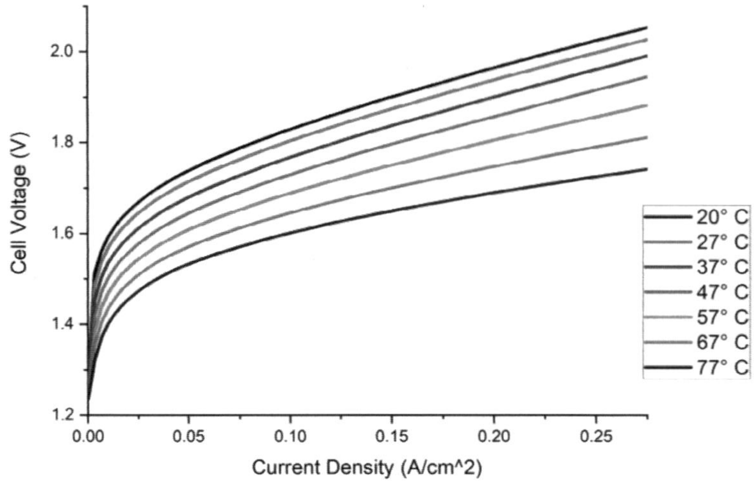

Figure 6. Variation of Cell voltage vs Current density plot with temperature at 7 bar Pressure

It can be observed that increase in temperature shifts the polarization curve downward toward lower voltage. The variation of V-I curve with change in temperature at constant pressure of 7 bar is shown in Figure 6. It can be observed that increase in temperature shifts the polarization curve downward toward lower voltage. As the temperature increases, the better electrode electrochemical performance is achieved, the conductivity of electrolyte increases and resistance of membrane decreases which in fact decrease the ohmic overpotential. Higher temperature also leads to higher rate of bubble departure which lead to decrease the effect of bubble. With the increase in temperature the all there defining parameters: reversible cell voltage, activation overpotential, and ohmic overpotential decreases hence overall cell voltage also decreases with increase in operating temperature.

3.2. *Effect of Pressure*

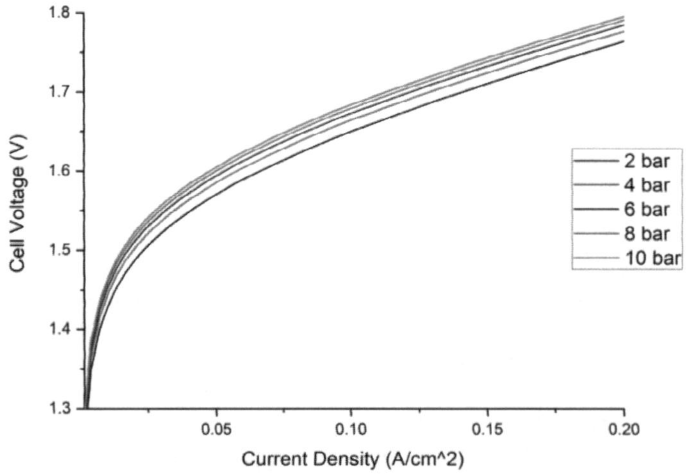

Figure 7. Variation with Pressure with 60° as operating temperature

The variation of polarization curve with change in operating pressure is illustrated in Figure 7. Pressure influences reversible voltage and bubble coverage factor. With increase in operating pressure the bubble coverage factor decreases hence activation overpotential and electrolyte resistance slightly decreases but is negligible. The effect of pressure is noticeable in reversible voltage. Reversible cell voltage increases non linearly with increase in pressure thus causing overall cell voltage to increase.

3.3. Effect of concentration

The concentration of KOH has a significantly effect on conductivity of the electrolyte hence on ohmic overpotential. The conductivity of KOH solution attains its peak value at around 30 wt% [33]. Increasing or decreasing the concentration from 30 wt% concentration, the conductivity of KOH decreases hence resistivity increases. Since ohmic overpotential show the significant effect on higher current density, the effect of change in concentration is also observable in higher current density. The effect of variation of electrolyte concentration on polarization curve is shown in Figure 7.

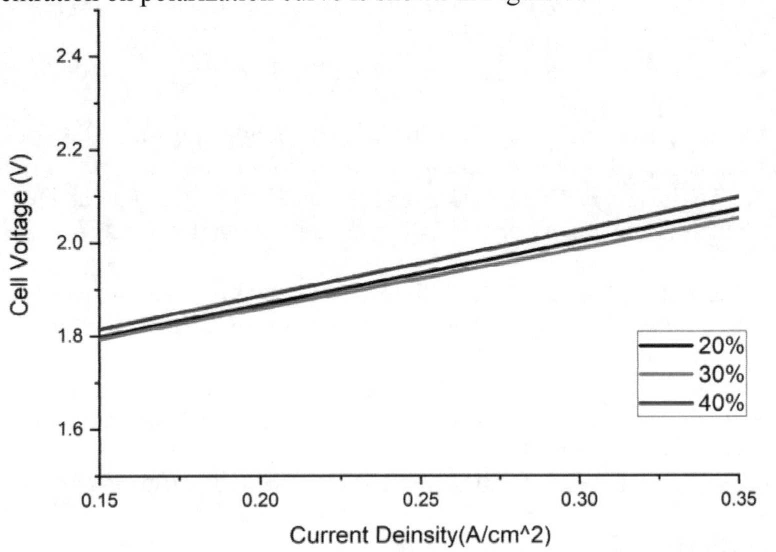

Figure 8. Variation of polarization curve with electrolyte concentration at 1 bar and 60°C

4. Conclusion

A mathematical model based on different thermodynamical, electrochemical and empirical equation with different input parameter has been developed in MATLAB/SIMULINK. Polarization curve based on the mathematical model has been generated and validated with previously published experimental data. The model was used to predict the behavior of polarization of on varying different input parameters. It was observed that increase in operational temperature and decrease in operational pressure, decreases the cell voltage of the electrolysis process. And 30 wt% of KOH solution as electrolyte exhibits the optimal cell voltage as compared to 20 and 40 wt% solution. With this model, effect of variation of different other parameters like thickness of electrode and membrane, roughness of electrode, porosity, tortuosity etc. on performance of electrolyzer can be studied. For further works, the model can be used to determine and compare the effect of variation of different parameters on efficiency of the electrolyzer and out gas flow rate of the electrolyzer.

References

[1] IEA 2021 Hydrogen *IEA*
[2] Idriss H, Scott M and Subramani V 2015 Introduction to hydrogen and its properties *Compendium of Hydrogen Energy* 3–19
[3] Nikolaidis P and Poullikkas A 2017 A comparative overview of hydrogen production processes *Renewable and Sustainable Energy Reviews* **67** 597–611
[4] International Energy Agency 2021 *Global Hydrogen Review 2021*
[5] Tijani A S, Yusup N A B and Rahim A H A 2014 Mathematical Modelling and Simulation Analysis of Advanced Alkaline Electrolyzer System for Hydrogen Production *Procedia Technology* **15** 798–806

[6] International Renewable Energy Agency 2020 Green Hydrogen Cost Reduction Scaling Up Electrolysers to Meet the 1.5°C Climate Goal

[7] Ulleberg I 2003 Modeling of advanced alkaline electrolyzers: a system simulation approach **28** 21–33

[8] Yakdehige Sanath Kumara De Silva 2017 Design of an Alkaline Electrolysis Stack

[9] Carmo M, Fritz D L, Mergel J and Stolten D 2013 A comprehensive review on PEM water electrolysis *Int J Hydrogen Energy* **38** 4901–34

[10] Amores E, Rodríguez J, Oviedo J and De Lucas-Consuegra A 2017 Development of an operation strategy for hydrogen production using solar PV energy based on fluid dynamic aspects *Open Engineering* **7** 141–52

[11] Sanath Y, De Silva K and Middleton P H 2017 Design of an Alkaline Electrolysis Stack TermoMag View project Roxsolidcell View project

[12] Falcão D S and Pinto A M F R 2020 A review on PEM electrolyzer modelling: Guidelines for beginners *J Clean Prod* **261**

[13] Maruf-ul-Karim M and Iqbal M T 2009 Dynamic modeling and simulation of alkaline type electrolyzers *Canadian Conference on Electrical and Computer Engineering* **2** 711–5

[14] Amores E, Rodríguez J and Carreras C 2014 Influence of operation parameters in the modeling of alkaline water electrolyzers for hydrogen production *Int J Hydrogen Energy* **39** 13063–78

[15] Corrêa J M, Farret F A, Canha L N and Simoes M G 2004 An electrochemical-based fuel-cell model suitable for electrical engineering automation approach *IEEE Transactions on Industrial Electronics* **51** 1103–12

[16] Mann R F, Amphlett J C, Hooper M A I, Jensen H M, Peppley B A and Roberge P R 2000 Development and application of a generalized steady-state electrochemical model for a PEM fuel cell *J Power Sources* **86** 173–80

[17] Adibi T, Sojoudi A and Saha S C 2022 Modeling of thermal performance of a commercial alkaline electrolyzer supplied with various electrical currents *International Journal of Thermofluids* **13** 100126

[18] Gilliam R J, Graydon J W, Kirk D W and Thorpe S J 2007 A review of specific conductivities of potassium hydroxide solutions for various concentrations and temperatures *Int J Hydrogen Energy* **32** 359–64

[19] Hammoudi M, Henao C, Agbossou K, Dubé Y and Doumbia M L 2012 New multi-physics approach for modelling and design of alkaline electrolyzers *Int J Hydrogen Energy* **37** 13895–913

[20] Olivier P, Bourasseau C and Bouamama P B 2017 Low-temperature electrolysis system modelling: A review *Renewable and Sustainable Energy Reviews* **78** 280–300

[21] Henao C, Agbossou K, Hammoudi M, Dubé Y and Cardenas A 2014 Simulation tool based on a physics model and an electrical analogy for an alkaline electrolyser *J Power Sources* **250** 58–67

[22] Shen M, Bennett N, Ding Y and Scott K 2011 A concise model for evaluating water electrolysis *Int J Hydrogen Energy* **36** 14335–41

[23] Janssen L J J, Sillen C W M P, Barendrecht E and van Stralen S J D 1984 Bubble behaviour during oxygen and hydrogen evolution at transparent electrodes in KOH solution *Electrochim Acta* **29** 633–42

[24] Vogt H and Balzer R J 2005 The bubble coverage of gas-evolving electrodes in stagnant electrolytes *Electrochim Acta* **50** 2073–9

[25] Denis Antipin M R 1967 Calculation of the Tafel slope and reaction order of the oxygen evolution reaction between pH 12 and pH 14 for the adsorbate mechanism **1** 5–24

[26] Abdin Z, Webb C J and Gray E M A 2017 Modelling and simulation of an alkaline electrolyser cell *Energy* **138** 316–31

[27] Phillips R, Edwards A, Rome B, Jones D R and Dunnill C W 2017 Minimising the ohmic resistance of an alkaline electrolysis cell through effective cell design *Int J Hydrogen Energy* **42** 23986–94

[28] Brauns J, Schönebeck J, Kraglund M R, Aili D, Hnát J, Žitka J, Mues W, Jensen J O, Bouzek K and Turek T 2021 Evaluation of Diaphragms and Membranes as Separators for Alkaline Water Electrolysis *J Electrochem Soc* **168** 014510

[29] Miles M H, Kissel G, Lu P W T and Srinivasan S 1976 Effect of Temperature on Electrode Kinetic Parameters for Hydrogen and Oxygen Evolution Reactions on Nickel Electrodes in Alkaline Solutions *J Electrochem Soc* **123** 332–6

[30] Maciej Matyka Tortuosity in porous media flow (interactive document)

[31] Gugliuzza A 2015 Membrane Wettability *Encyclopedia of Membranes* 1–4

[32] Rodríguez J and Amores E 2020 modeling and experimental validation of an alkaline water electrolysis cell for hydrogen production *Processes* **8** 1–17

[33] Brauns J and Turek T 2020 Alkaline water electrolysis powered by renewable energy: A review *Processes* **8**

Simulation and Field-Based Study of Thermal Performance of a Building - A Case Study

S Malesh[1*], G Kamal[1], J Yaman[1] and B Bivek[1]
[1]Kathmandu University, Dhulikhel, Nepal
*Corresponding Author: malesh.shah@ku.edu.np

Abstract. Concerns about building thermal performance and energy consumption are fast growing as they account for roughly 40% of worldwide annual energy use. Building construction in Nepal is controlled by the National Building Codes (NBC), which focus solely on structural considerations and ignore thermal behaviour. As a result, housing tends to be too much warm and too cold during summer and winter respectively due to the lack of passive heating and cooling mechanisms considered during the construction phase. The Thakle Integrated Settle (TIS), built by the Department of Urban Development and Building Construction (DUDBC) under the National Reconstruction Authority and supported by Oxfam Nepal, was the subject of this research. The building's performance in the chosen location was analysed using a field and TRNSY simulation-based technique. Using meteorological data from Meteonorm software, the TRNSYS model was created to examine the performance in terms of internal air temperature and heating/cooling load of two buildings with different orientations, e.g., North-East (NE) and South-West (SW). According to simulation results, there is a heating demand that covers roughly 37% of the year, while cooling and comfort conditions were 32% and 31%, respectively. For NE and SW-facing buildings, heating demand was 4.10 MWh and 4.07 MWh, while cooling demand was 1.302 MWh and 1.309 MWh respectively.

Keyword. TRNSYS, Active Retrofitting, Building Energy, Simulation, Thermal Comfort

1. Introduction

Nepal is divided into four distinct bioclimatic zones based on elevation above sea level as warm, temperate, cool, and cold temperatures [1]. Depending on the climatic zone, the annual maximum average temperature ranges from 18°C to 36°C while the minimum ranges from 6°C to 18°C [2]. Also, Nepal is a diverse country in terms of geography, culture, socioeconomics, climate, and other factors. Based on this housing styles and building materials vary from Terai to Himalayan areas depending on the local environment or climate, locally accessible resources, and regional development. These structures are constructed using old traditional and cultural architecture. However, as a result of economic progress and modernization, individuals began to embrace western home designs and contemporary materials while building, resulting in the rise of various modern houses during the previous ten years. This number is only expected to rise in the future. Even though modern structures are created with the most up-to-date materials, they still lack basic design standards for building performance and thermal comfort aspects such as insulation, airtightness, and ventilation.

NBC governs the construction of structures in Nepal. Recently after the devastating earthquake in 2015, the DUDBC of the Ministry of Urban Development of the Government of Nepal is more concerned about the enhancement of the design requirement in terms of structural aspects to lessen the damage caused by seismic risks [3]. There are altogether 25 NBC to date with safety as the top consideration, none of these has provision for housing thermal comfort and energy-efficient building.

IOP Conf. Series: Materials Science and Engineering 1279 (2023) 012006 doi:10.1088/1757-899X/1279/1/012006

On the other hand, various studies suggest that historic buildings, such as those constructed with mud mortar and wooden joists, have greater thermal efficiency than contemporary buildings, such as those constructed with cement mortar, steel structures, and concrete floors [4]. The traditional structure was 1 to 2°C warmers in the winter and 1 to 2°C coolers in the summer than modern residential buildings, saving a minimum of 10-20% energy for either heating or cooling [5].

Most of the available building energy related studies for Nepal are not site specific. This work presents the status of thermal situations at a specific community. TRNSYS was used to visualize the indoor air temperature, heating, and cooling load of two identical building structures with variation in orientation with the incorporation of data obtained from the field survey.

2. Material and Methods

The TIS is located in Sindhupalchok District which was built in collaboration with the local government, DUDBC and Oxfam Nepal. A survey conducted to investigate the status of thermal comfort was corroborated with simulations. The simulation tools used namely TRNSYS18 to perform energy modelling of the building and Meteonorm 8 to obtain the weather file for the given location. Initially, the construction of the building has been designed in the subprogram of TRNSYS called TRNBuild, with measured data from the field as input which includes construction materials, dimensions, and orientation of walls, ceiling, floor, doors, windows, and roof. The detail of the building material with layer-wise construction used in the studied buildings are listed in Table 1 with their U-values obtained from TRNSYS.

Table 1. Detail of Building Materials.

Nature	Composition	Thickness (m)	U value [W/m^{2}K]
External Walls	Plaster	0.210	2.824
	Brick Block		
Internal Walls	Plaster	0.220	2.768
	Brick Block		
	Plaster		
Ceiling	Wooden Material	0.02	3.472
Floor	Concrete Floor	0.100	4.595
Roof	GI Sheet	Massless	5.878
Window	Single Glazed	Area: 0.86	5.68
Door	Plywood	0.02	3.297

The overall floor area of all the 31 identical buildings in the TIS were 42.60 m^2 with sixteen of them facing NE and the remaining fifteen facing SW direction. In TRNBuild total of six thermal zones have been created which indicate '*KITCHEN*', '*ROOM1*', '*ROOM2*', '*ROOM3*', '*BATHROOM*', and '*ATTIC SPACE*'. The analysis house has been considered to have a residence of 5 people with a real-time schedule of their presence in the room and activity level defined using Type14h. The zone has constant infiltration of 0.5ACH as per ASHRAE [6], a lighting power of 10W, and a Television of 100 W with their schedule defined for individual zone. The TRNSYS simulation model consists of many components, including weather data (Type15-2) which was obtained from Meteonorm in TMY2 format, multizone building (Type56), integrators (Type24), forcing functions (Type14h), soil temperature profile (Type77), printer (Type25-c) and equations linked together [7]. A schematic layout of the developed model is presented in Figure 1.

The indoor zone air temperature was calculated with heating and cooling medium off. Then heating and cooling set point temperatures has been set to 20°C and 26°C respectively for the heating and cooling load of the analysis building. Heating and cooling will be activated only when there are occupants in the room and zone temperatures fall below the heating and cooling setpoint temperature. The same steps have been repeated for SW-oriented buildings too.

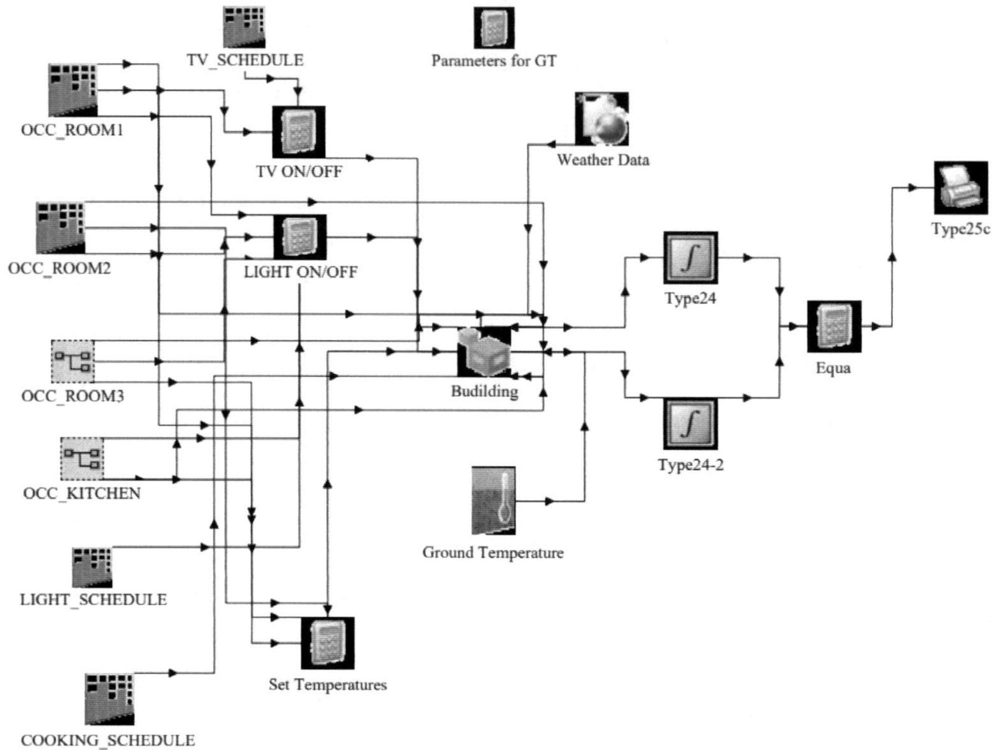

Figure 1. TRNSYS Model for the Building Analysis

3. Result and Discussions

It was discovered from the survey that the inhabitants experience extreme thermal environment at different time of the day and similar goes for different seasons. Simulation was done to observe the thermal performance of the buildings using TRNSYS with a time step of one hour for one year. The inbuilt output functions of TRNSYS specified by NType30 (QHEAT), NType31 (QCOOL), and NType1 (TAIR) were used to obtain the required data which were further processed for analysis.

3.1. Air Node Temperature

The year-round hourly temperature of all thermal zones for both NE and SW-facing buildings depicts higher temperatures in the zone having higher solar incidence radiation on the walls in terms of both intensity and exposure time. This is illustrated in Figures 2(a) and (b) from the period of 745 to 1416 hours (i.e., February). In the case of NE, the temperature is higher in *ROOM2* which receives daylight solar radiation while it contrasts in the case of SW. The same pattern exists for the kitchen as *ROOM1* and *ROOM3* align with *ROOM2*. The maximum temperature is around 32°C in summer and a minimum of 6°C which are certain temperatures higher and lower compared to ambient air temperature respectively. Further, annual air node temperatures of all four thermal zones have been divided into three categories: heating degree hours (20°C), cooling degree hours (26°C), and thermally comfort zone ($20°C \leq T \leq 26°C$) for both the NE and SW faced buildings.

The categorization of temperatures in different heating degrees for NE and SW is shown in Figures 3 (a) and (b). This shows that heating periods are more dominant in both buildings with more requirements in SW. Also, the orientation of the zone within the same building plays a significant role in heating and cooling requirements. Overall, the average heating, cooling, and thermal comfort fraction are 37, 32, and 31 percent respectively for both orientation buildings.

IOP Conf. Series: Materials Science and Engineering 1279 (2023) 012006 doi:10.1088/1757-899X/1279/1/012006

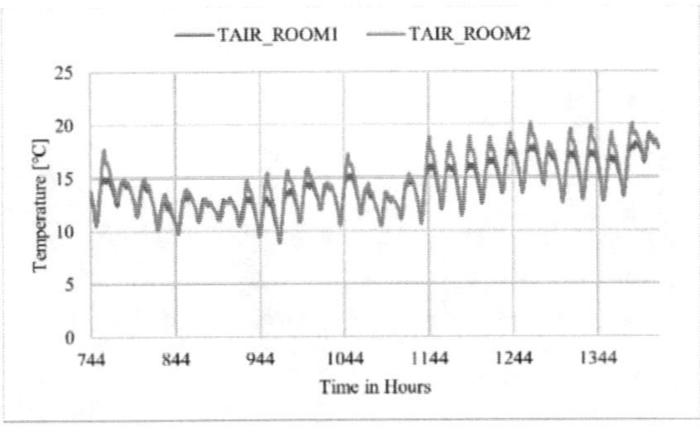

(a)

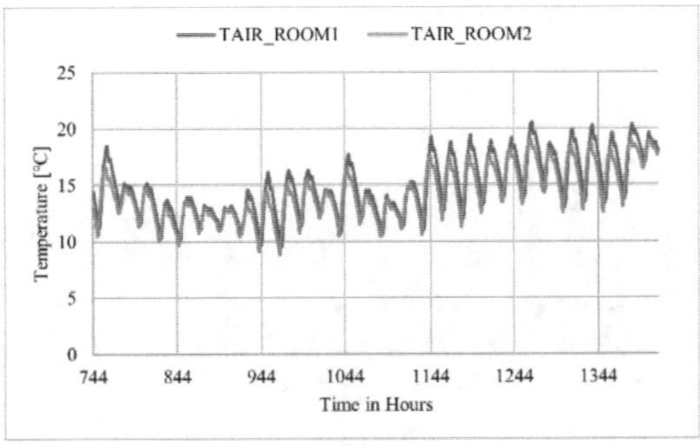

(b)

Figure 2. (a) Indoor air temperature of NE-faced building for February (b) Indoor air temperature of SW-faced building for February

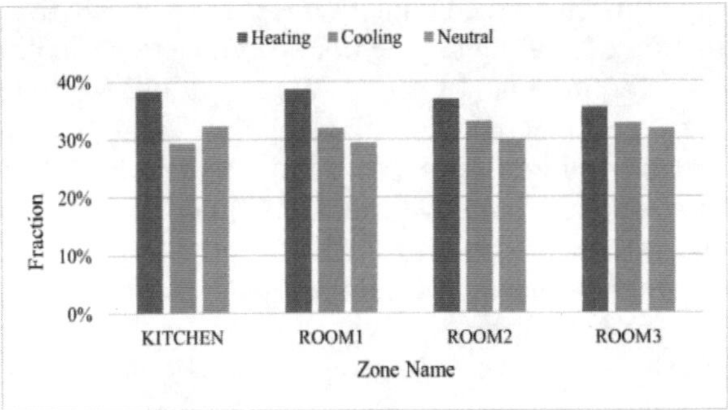

(a)

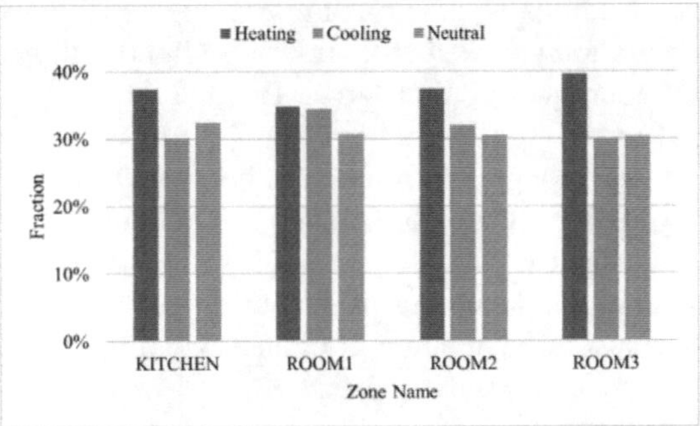

(b)

Figure 3. (a) Heating Distribution for NE Building (b) Temperature Distribution for SW Building

3.2. Heating and Cooling Load

This is an alternative method of evaluating indoor thermal comfort or quantifying the energy demand for heating and cooling of the building [8]. Annual energy demand was calculated on an hourly basis for all four zones of NE and SW-facing buildings which were further integrated to obtain the monthly requirement. The monthly heating and cooling requirements for the two-orientation building are depicted in Figures 4 (a) and (b). It shows that heating is considerable during January – March, and November – December for both structures with higher in January followed by December and February. While cooling is throughout the summer season with minimal in April and October. However, the cooling load is not significant even though they tend to appear for larger months in a year. For an identical building with a different orientation, the heating demand of the NE building was slightly greater than the cooling requirement of the SW building. Thus, orientation has an impact on zone temperature and, consequently, on a building's heating and cooling demand.

Overall, the heating and cooling patterns were dependent on the temperature difference between the set point temperature, room air temperature, and ambient air temperature. The higher temperature on the positive side i.e., higher ambient temperature requires cooling and the magnitude is directly proportional to the temperature rise. On the other hand, temperature differences on the tolerance zone have zero heating and cooling loads while lower indoor temperature compared to the set point and the ambient temperature requires the heating load whose magnitude again depends on the temperature range.

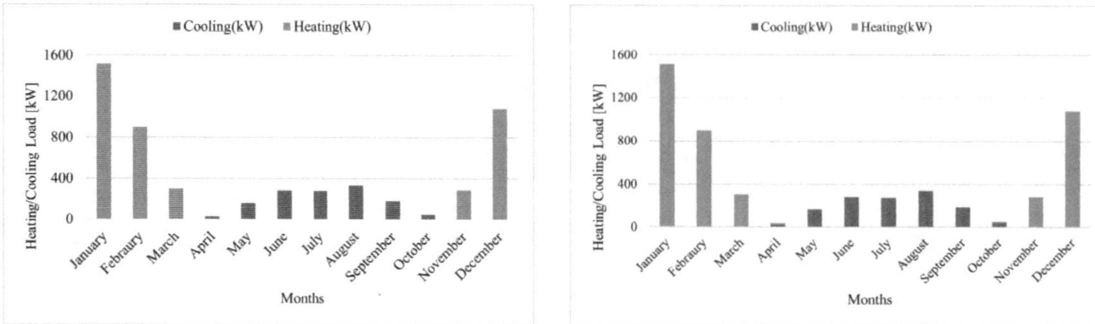

(a) (b)

Figure 4. (a) Heating and Cooling Load for NE Building (b) Heating and Cooling Load for SW Building

4. Conclusion

The findings obtained from literature and survey are in line with the simulation results. The TRNSYS simulation shows that there was a slightly higher fraction of heating degree hours compared to cooling and thermally comfort zone. However, the heating and cooling load of 4.10 MWh and 1.302 MWh for NE and 4.07 MWh and 1.309 MWh for SW for a year suggest that there is higher heating demand despite having a not greater fraction of heating days compared to cooling. In addition, a room that receives mid-day sunlight tends to be much hotter than the other two rooms. Even the orientation plays a part in the indoor thermal comfort of houses. As a result, this work suggests that it is vital for NBC to care about the thermal behaviour of the buildings, which will produce a comfortable climate inside the homes and save a significant amount of energy with small construction modifications.

5. References

[1] Bodach S, developing bioclimatic zones and passive solar design strategies for Nepal, 30th Int. PLEA Conf. Sustain. Habitat Dev. Soc. Choose. W. Forw. - Proc., 1 December 2014; 167.

[2] DHM, Study of Climate and climatic variation over Nepal Submitted by: Nepal Hydrological and Meteorological, 2015; 41.

[3] National Reconstruction Authority, Seismic Design of Buildings in Nepal, Ministry Urban Development, 2020; 1.

[4] Khadka B. Mud masonry houses in Nepal: A detailed study based on entire reconstruction scenario in 31 earthquake-affected districts. Structures. 2020;25:816–38.

[5] Chenvidyakarn T. Passive design for thermal comfort in hot humid climates. Journal of Architectural/Planning Research and Studies (JARS). 2018;5(1):1–28.

[6] ANSI/ASHRAE 55:2004 Thermal Environmental Conditions for Human Occupancy, ASHRAE, 2004; 30.

[7] Ahamed MS, Guo H, Tanino K. Modeling heating demands in a Chinese-style solar greenhouse using the Transient Building Energy Simulation Model TRNSYS. Journal of Building Engineering. 2020;29:101114.

[8] Monstvilas E, Stankevicius V, Karbauskaitė J, Burlingis A, Banionis K. Hourly calculation method of building energy demand for space heating and cooling based on steady-state heat balance equations. Journal of Civil Engineering and Management. 2012;18(3):356–68.

Acknowledgments

This research was supported by the Energy Systems and Technology Research Laboratory (ESTRL), Kathmandu University in terms of TRNSYS and Meteonorm License and all the supervisions. The funding for the research was provided jointly through University Grant Commission (UGC) Nepal and Energize Nepal Programme, Kathmandu University.

Techno-Economic Analysis of Energy Systems in Thakle Namuna Basti - A Case Study

Malesh Shah[1*], Kshitiz Koirala[1], Saugat Upreti[1] and Nawaraj Sanjel[1]
[1]Kathmandu University, Dhulikhel, Nepal
*Corresponding author: malesh.shah@ku.edu.np

Abstract. The electrical supply system in most underdeveloped nations is incredibly undependable. The country's existing distribution system has experienced regular power outages due to population growth, industrial expansion, rough terrain, and a long-distance transmission network. The conventional approach to centralized generation and transmission is failing to address this challenge and as a response, the utilization of renewable energy sources is expanding day by day to alleviate the energy crisis. The purpose of this study is to perform a techno-economic analysis of a 100% renewable off-grid and on-grid hybrid energy system for the electrification of Thakle Namuna Basti, a small communal village in Melamchi, Sindhupalchwok. The HOMER Pro 3.14 program is used to analyze the feasibility of the suggested hybrid power system. The available resources are provided as an input parameter along with load data and HOMER Pro optimizes the list of system architecture for that specific site. In addition, sensitivity analysis is performed for the proposed HES (Hybrid Energy System). The main outcome of the project is to find out the best system architecture based on various economic parameters such as NPC (Net Present Cost), LCOE (Levelized cost of electricity) and Operating and Maintenance Cost. The system having lowest NPC and LCOE makes the best system architecture.

Keywords: Net Present Cost, Levelized Cost of Electricity, Department of Urban Development and Building Construction, HOMER Pro, Internal Rate of Return

1. Introduction

Energy is a critical component of a developing country's rural socio-economic development. To achieve economic growth, poverty eradication, employment creation, and village livelihood improvement, rural electrification is crucial [1]. At present, developing countries lack grid-connected electricity. Grid extension is impractical due to dispersed populations and rugged terrain. Thus, as a result, small off-grid standalone renewable energy systems are a viable option for ending the electricity gap in rural areas of developing countries, where grid extension is lagging behind population growth. Even though these small-scale energy systems generate relatively little electricity, they can make a significant contribution to life quality and human sustainability by improving net energy production, public health and education for rural households in developing countries [2].

Due to advancements in technology, enabling regulatory frameworks in solar home systems and successful off-grid electrification projects have been established in numerous nations utilizing technologies such as mini-grids and solar household systems. Additionally, the growing need to reduce carbon emissions from fossil fuel-based networks, the rising cost of grid expansion, the declining cost of renewable energy alternatives while improving performance and the inaccessibility of various settlements all favored the distributed generation approach [3]. Hence the solution for this situation could be to promote alternative forms of clean and eco-friendly energy generation, such as solar, wind, hydro, biogas, biomass, tidal, wave, ocean thermal energy, etc. [4].

In the developed world, mini-grids are used to increase energy security, power quality and stability as well as to avoid blackouts during natural disasters. Reduce the loss of energy during transmission and distribution. However, in developing countries, mini- grids are used to increase energy access for isolated rural and island populations. Island settlements, whether developed or developing also use the concept of mini-grids [3].

The most common approaches to decentralized electrification are stand-alone with local generation, a pure extension of the central grid and hybrid solutions. Depending on the technical and financial feasibility, one of the models can be used; however, until now, in most of the cases solutions have been limited to either a pure extension of the central power grid or a stand-alone micro grid. The primary advantage of hybrid systems is that when a variety of energy sources are combined, the system's reliability improves. Additionally, there are cost reductions associated with renewable energy technologies as a result of R&D, and increased deployment makes them more appealing to investors [5].

Global gross electricity production increased by 3.9 percent in 2018 over 2017. Since 1974, except in 2008 and 2009, global electricity production has increased at an annual rate of 3% [6]. Similarly, in Nepal's fiscal year 2019/20, a total of 3021 GWh of electricity is generated, an increase of 18.57 percent over the 2548 GWh generated in FY 2018/19. However, insufficient production means that 1,729 GWh of electricity is imported from India in 2019/20. The majority of electricity is generated by hydropower plants while 53.41 MW is generated by thermal power plants and 1.350 MW is generated by solar power plants [7]. Enhancements to the industrial and service sectors electricity supply can contribute to increased economic growth. Growth in the electricity sector is expected to continue, owing to rapid urbanization and an increase in the number of energy-intensive industries. Electricity demand is expected to increase from 10,138 GWh in 2019 – 2020 to 31,196 GWh in 2029 – 2030 [8].

Exploring renewable energy sources eliminates the need for electricity transmission to remote rural areas, raises living standards and promotes a clean environment. These non-traditional or renewable energy sources can meet demand and, in some cases, exceed it without posing environmental risks. As a result, this option is more suitable for addressing the issue of rural energy requirements, as these sources are abundant in rural areas. This decentralized energy generation will contribute to resolving the rural electrification crisis. The combustion of wood and agricultural waste for heating and cooking is inefficient and contributes to pollution through the production of gases such as CO2 [9]. To devise an effective solution to this problem, a thorough assessment of the current state of affairs must be made. The plan that will be developed will depend on how the village's existing resources are utilized and how they can be converted into useful and clean forms of energy for village households.

The disadvantages of wind and solar energy are their unpredictability and reliance on weather and climate change. Additionally, wind and solar energy variations may not correspond to the time distribution of electricity demand. In such cases, decentralized renewable hybrid power can provide viable alternatives, wind/photovoltaic/battery, wind/battery photovoltaic/battery, and other solutions without a battery backup option [10]. Each renewable energy system must undergo a techno-economic analysis to determine its efficiency and economic viability. The analysis is complicated by the fact that it involves a large number of generating systems. Additionally, the optimal design capacity of energy facilities is highly dependent on electric demand and the extent to which renewable energy is consumed, both of which are difficult to forecast in advance. To avoid excessive energy consumption, design capacity must be calculated in conjunction with operation planning, which is based on anticipated energy demand and production. Any additional energy generated as a result of inefficient capacity planning and operation is almost certainly wasted [11]. The HOMER Pro program could easily be used to simulate and determine the optimal component arrangement for the case study. We investigated off-grid and grid-connected systems with and without battery storage. The financial and environmental implications of several cases are then discussed, including the COE (Cost of Electricity), NPV (Net Present Value), IRR (Internal Rate of Return), renewable fraction, CO_2 emissions, etc.

2. Study Area

The study area is Thakle Namuna Basti which lies around 5 to 6 km away from Bahunepati, Melamchi and is located in Melamchi Municipality of Sindhupalchwok District with longitude and latitude of 27 48'14" N 85 32'49" E. There are around 130 people in that community and the total number of houses is 32. Each house is identical as it was, built by the DUDBC (Department of Urban Development and Building Construction) under the National Reconstruction Authority and supported by Oxfam Nepal. The primary source of income and livelihood is agriculture whereas youth work as drivers, technicians and so on. The majority of people belong to Tamang tribes. The village makes use of both traditional and modern energy sources. The traditional energy sources are fuelwood and crop residue etc. for cooking. Electricity is used for lighting and few people use LPG for cooking purposes. Most of the family members especially women spend a considerable amount of time collecting fuelwood. Most of the electricity is consumed for lighting though some people even use television, induction stove, water boiler, etc. The figure 1 shows the google view of Thakle Namuna Basti [12].

Figure 1. Thakle Namuna Basti from google maps

3. Load Profile of the Study Area

A survey was conducted to find the load of the study area. Customer's electricity bills were used to find the monthly load which was compared with load demand to get an idea of exact load demand of that place [13]. Studies have revealed that a large proportion of the variability in electricity demand is dependent on weather variables such as air temperature, humidity, wind speed, cloud cover, and luminosity [14]. So, to find the variability on the load profile the result of the survey, considering the appliances that are used and their operating hours and even the load profile of the country were taken in reference. Figure 2 shows the monthly load profile of Thakle Namuna Basti given by Homer Pro.

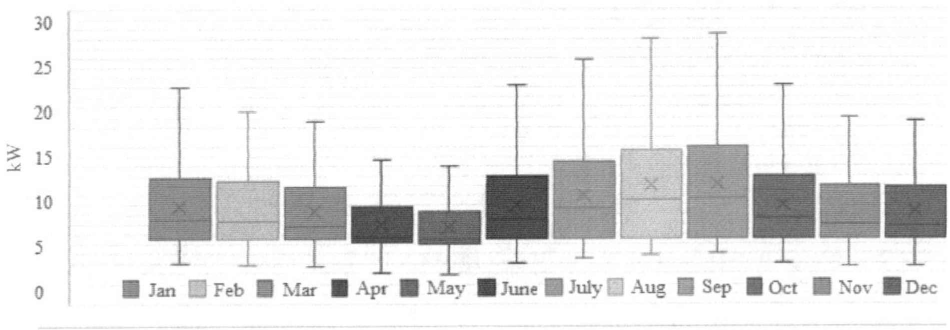

Figure 2. Monthly load profile of Thakle Namuna Basti

4. Resources Available

After the field visit, data is collected from different sources and studied. The potential of available renewable energy resources such as solar energy, wind energy, and hydropower were considered.

4.1. Solar Potential

For solar radiation, National Renewable Energy Lab Database [15] is used. It was found out that the average solar radiation in the selected area is 4.97 kWh/m^2/day similar data from the Meteonorm. Data from Meteonorm is also taken and after comparative study, National Renewable Energy Lab data is used as there is not much difference in average monthly irradiance. The curve in figure 3 shows the monthly average solar global horizontal irradiance and the bar graph in figure 3 shows daily radiation of the study area of every month.

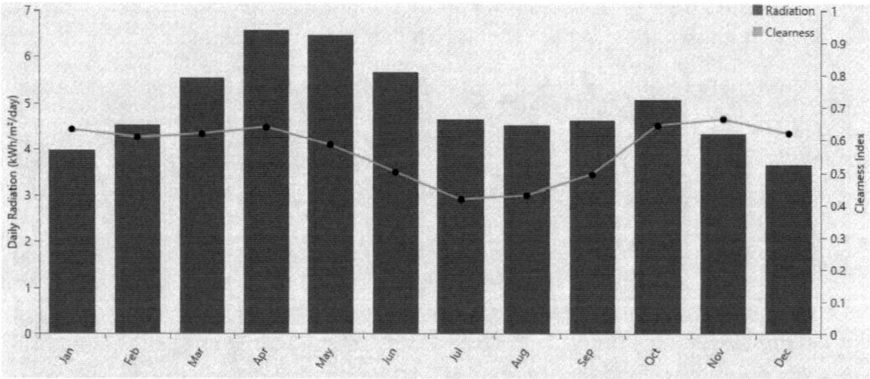

Figure 3. Daily Radiation along with Monthly average solar data

4.2. Wind Potential

For wind resources, HOMER Pro estimate data from NASA Surface meteorology and solar energy database [15] is used which gave the monthly average wind speed at 50 meters above the surface of the earth over 10 years' time period (July 1983 – June 1993). The figure 4 shows the average monthly wind speed of every month for 10 years from the HOMER Pro Database.

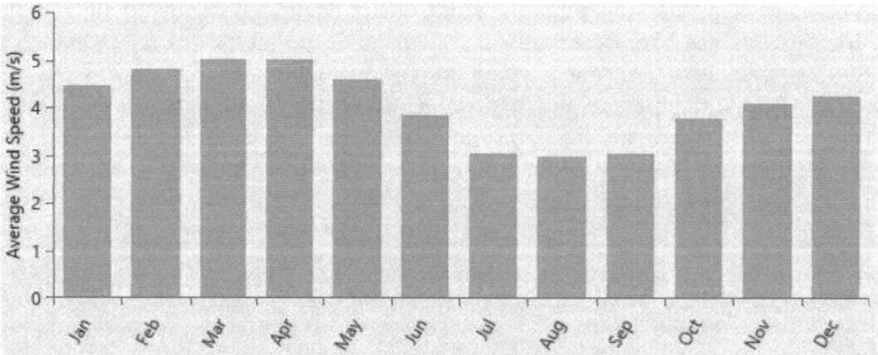

Figure 4. Monthly average wind speed data

4.3. Temperature Variation

The temperature data file from the Meteonorm is imported into HOMER Pro. The below figure shows the temperature of Thakle Namuna Basti in each month for a year. The lowest average monthly temperature is 9.9 °C in January and the highest average monthly temperature is 23.8 °C in June, July, and Aug and the average annual temperature is 18 °C.

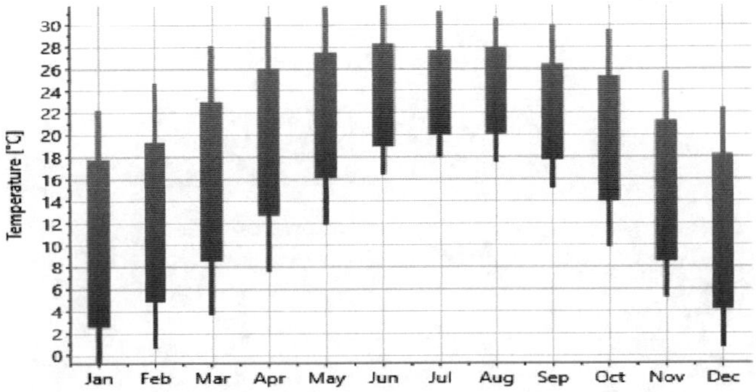

Figure 5. Monthly average temperature variation data

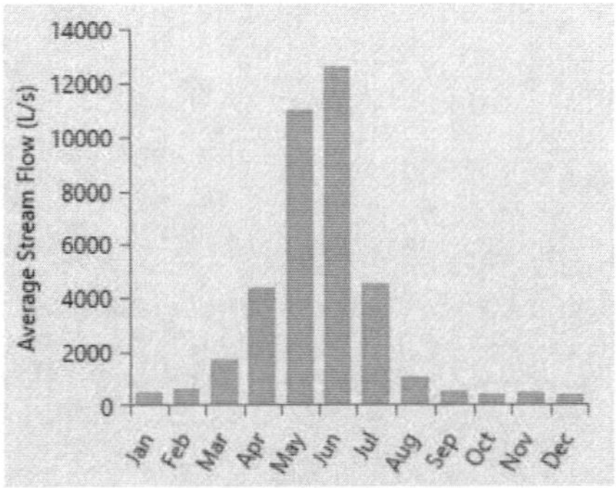

Figure 6. Monthly average flow rate of Sindhu River

4.4. Nearby Hydro Resource

The hydro resource that is available in the study area is the Sindhu River. For calculating the power, the flow rate and head of the river is required.

$$P = \rho \, Q \, g \, H$$

Where,

ρ = Density of water

Q = Flow rate of water

g = Acceleration due to gravity

H = Head from turbine to reservoir

To find the flow rate Stream- flow Prediction Tool which is provided by ICIMOD [16] is used which gave an estimated flow rate of the river in time series.

5. System Components

5.1. Proposed system architecture in HOMER Pro

Two system architecture were used for analysis. One with grid and another without grid. Figure 7 (a) shows the system architecture for off grid system and figure 7 (b) shows system architecture for an on-grid system.

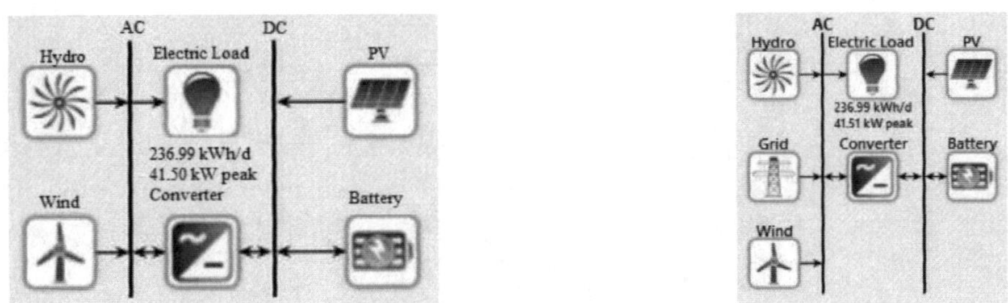

(a) (b)

Figure 7.(a) System architecture (Off grid system) Figure **7.**(b) System architecture (On-grid system)

5.2. Wind Turbine

A generic wind turbine model is considered for this study. Its nominal power is 10 kW at 13 m/s. The technical specifications of the turbine are presented in the table 1, while its power curve is also shown in the figure 6 which shows the variation of the output power to the wind speed where the output is maximum at a rated speed of 13m/s [17].

Table 1. Specification of Wind Turbine

Items	Specification
Model	Generic 10 kW
Manufacture	Generic
Rated Capacity	10 KW @13 m/s
Hub Height	24 m
Output type	AC
Life Time	20 Years
Cut-in Wind Speed	3 m/s
Max. Design Wind Speed	60 m/s

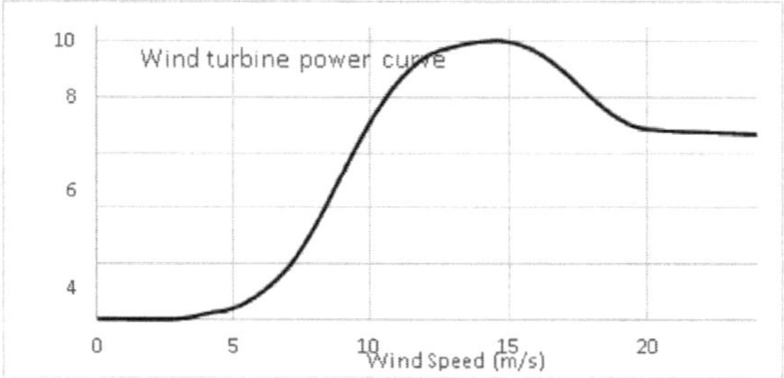

Figure 8. Wind turbine power curve

5.3. Photovoltaic Arrays

1 kW Generic flat-plate PV is selected for this project. A dedicated inverter technical specification for this type of PV array is presented in Table 2. Considering the market for renewable energy sources in Nepal, the installation, replacement, and maintenance costs for a 1 kW solar energy system are estimated to be \$1100, \$1100, and \$10 per year respectively [18].

Table 2. Specifications of PV Arrays

Parameter	Value
Rated Capacity	1 KW
Derating Factor	80%
Nominal operating cell temperature	47.5 C
Ground reflectance	20%
Lifetime	25 years

5.4. Battery

EnerSys PowerSafe SBS 1200 is selected for the study. The cost for the battery is considered \$646.10 and the replacement cost is \$646.10. The operation and maintenance cost of the battery is considered as \$5 per year with a life span of 15 years [19].

Table 3. Specification of Battery

Items	Specification
Model Name	EnerSys PowerSafe SBS 1200
Manufacturer	EnerSys
Type	Lead Acid Battery
Model	Kinetic
Capacity(kWh)	16.68
Duration (hours)	0.6
Max charge current (A)	1.20E+03
Max discharge current(A)	2.30E+03
Capacity Ratio	0.297
Roundtrip efficiency (%)	97

5.5. Converter

A power electronic converter is required for maintaining the energy flow between the AC and DC buses. Installation and replacement costs for a 1 kW converter system were estimated to be $300 in both cases. The lifetime of a unit was estimated to be 15 years with an efficiency of 95 % [19].

6. Results for On-Grid System

6.1. Sensitivity Analysis

Table 4 shows the list of sensitivity variables that are used during this simulation. Total four sensitivity variables were taken which were Power price, Sellback rate, Hydropower cost and Hydro flow rate.

Table 4. List of sensitivity variables for on-grid system

Power Price($/kWh)	Sellback Rate($/kWh)	Hydropower Cost($/kWh)	Hydro flow rate(L/s)
0.14	0.04	15000	50
0.15	0.05	18000	75
0.16	0.06	20000	100
	0.07	22000	
		25000	

Initial condition:
Power Price - 0.14 $ Hydro Capital Cost – 15000 $
Hydro Flow rate – 75 L/s Sellback Rate - 0.04 $
The table 5 shows the results of the base case and lowest case system while changing the power price at the initial condition.

Table 5. Sensitivity analysis when power price is varied in on-grid system

Power Price($/kWh)	Parameters	Base Case System	Lowest Case System
0.14	NPC	$76,591	$70,507
	Operating Cost	$4764/yr	$2,909/yr
	LCOE	$0.0623/kWh	$0.0588/kWh
0.15	NPC	$80,384	$71,829
	Operating Cost	$5,058/yr	$2,981/yr
	LCOE	$0.0654/kWh	$0.598/kWh
0.16	NPC	$84,176	$72,922
	Operating Cost	$5,351/yr	$2,419/yr
	LCOE	$0.0685/kWh	$0.0588/kWh

Table 6. Sensitivity analysis when sellback rate is varied in on-grid system

Sellback Rate($/kWh)	Parameters	Base Case System	Lowest Case System
0.04	NPC	$76,591	$70,507
	Operating Cost	$4,764/yr	$2,909/yr
	LCOE	$0.0623/kWh	$0.588/kWh
0.05	NPC	$75,482	$71,269
	Operating Cost	$4,679/yr	$3,066/yr
	LCOE	$0.0614/kWh	$0.598/kWh
0.06	NPC	$74,373	$72,072
	Operating Cost	$4,593/yr	$3,244/yr
	LCOE	$0.0605/kWh	$0.0591/kWh
0.07	NPC	$73,264	$66,205
	Operating Cost	$4,507/yr	$-14,849/yr
	LCOE	$0.0596/kWh	$0.0142/kWh

IOP Conf. Series: Materials Science and Engineering 1279 (2023) 012007 doi:10.1088/1757-899X/1279/1/012007

From table 5, we can conclude that when power price is increased then NPC, operating cost and LCOE increase, but LCOE decreases at power price of $0.16 at lowest case system. The table 6 shows the results of the base case and lowest case system while changing the sellback rate at the initial condition. From table 6, we can conclude that when sellback rate is increased then NPC, Operating cost and LCOE decreases. Table 7 shows the results of the base case and lowest case system while changing the Hydropower cost at the initial condition. From table 7, we can conclude that when hydropower cost is increased then NPC and LCOE increases but operating cost remains constant. The table 8 shows the results of the base case and lowest case system while changing the Hydro flow rate at the initial condition.

Table 7. Sensitivity analysis when hydropower cost is varied in on-grid system

Hydropower Cost($/kWh)	Parameters	Base Case System	Lowest Case System
15000	NPC	$76,591	$70,507
	Operating Cost	$4,764/yr	$2,909/yr
	LCOE	$0.0623/kWh	$0.0588/kWh
18000	NPC	$79,591	$73,455
	Operating Cost	$4,764/yr	$2,906/yr
	LCOE	$0.0648/kWh	$0.0614/kWh
20000	NPC	$81,591	$75,455
	Operating Cost	$4,764/yr	$2,906/yr
	LCOE	$0.0664/kWh	$0.0631/kWh
22000	NPC	$83,591	$77,455
	Operating Cost	$4,764/yr	$2,906/yr
	LCOE	$0.0680/kWh	$0.0648/kWh
25000	NPC	$86,591	$80,455
	Operating Cost	$4,764/yr	$2,906/yr
	LCOE	$0.0704/kWh	$0.0673/kWh

Table 8. Sensitivity analysis when flow rate is varied in on-grid system

Hydro flow rate(L/s)	Parameters	Base Case System	Lowest Case System
100	NPC	$52,345	$47,411
	Operating Cost	$2,889/yr	$2,236/yr
	LCOE	$0.0378/kWh	$0.0393/kWh
75	NPC	$76,591	$70,507
	Operating Cost	$4,764/yr	$2,909/yr
	LCOE	$0.0623/kWh	$0.0588/kWh
50	NPC	$106,435	
	Operating Cost	$6,054/yr	
	LCOE	$0.08421/kWh	

From table 8, we can conclude that when the hydro flow rate is decreased then NPC, LCOE and operating cost increases. At 50 L/s the base system wins so there is no separate base case and lowest case system.

6.2. List of all the system architecture

A total of 14 system architectures was proposed by Homer Pro. Based on the cost of electricity, the system architecture having hydro and grid comes out at the lowest price and similarly for lowest NPC the system having 4 batteries, 7.64 kW of converter along with grid and hydro is proposed. Also, for the lowest operating cost, the system having 1.82 kW of Photovoltaic, 3 batteries, 509 kW of converter along with hydro and grid is proposed.

The figure 7 also shows that the system with turbines has the highest LCOE, NPC and operating cost and the system with grid and hydro has the lowest LCOE, NPC and operating cost.

Architecture								Cost			
PV (kW)	Wind	Battery	Grid (kW)	Hydro (kW)	Converter (kW)	Dispatch		NPC ($)	COE ($)	Operating cost ($/yr)	Initial capital ($)
10.9		5	999,999	5.89	8.00	CC		$70,507	$0.0588	$2,909	$32,895
		2	999,999	5.89	2.36	CC		$74,652	$0.0652	$4,492	$16,578
			999,999	5.89		CC		$76,591	$0.0623	$4,764	$15,000
0.316			999,999	5.89	0.322	CC		$76,693	$0.0622	$4,740	$15,414
12.2	1	6	999,999	5.89	10.4	CC		$146,870	$0.118	$3,667	$99,467
	1	2	999,999	5.89	2.96	CC		$149,338	$0.129	$5,313	$80,649
	1		999,999	5.89		CC		$151,770	$0.120	$5,629	$79,000
2.47	1		999,999	5.89	1.31	CC		$152,211	$0.118	$5,412	$82,248
59.0			999,999		44.7	LF		$157,195	$0.0813	$5,992	$79,738
59.9		1	999,999		45.5	CC		$158,542	$0.0813	$5,952	$81,593
55.5	1		999,999		46.9	LF		$230,940	$0.121	$7,062	$139,640
54.8	1	1	999,999		47.0	CC		$231,647	$0.122	$7,133	$139,429
	18	20	999,999		23.0	CC		$1.52M	$1.03	$27,098	$1.17M
	22		999,999			CC		$1.82M	$0.955	$31,764	$1.41M

Figure 9. List of all system architecture proposed by HOMER Pro for on-grid system

6.3. System architecture at the initial condition

Two types of system architecture were proposed by HOMER Pro. One is the winning system architecture which is the system architecture with the lowest NPC. At initial condition, the winning system architecture has hydro of 5.89 kW, converter of 8 kW and PV of 10.9 kW along with the grid. But when the sensitivity variables are changed then the winning system architecture also changes and if there are no sensitivity variables then this section will appear blank.

Another one is base case architecture which is the system with the lowest initial capital cost. At initial condition, base case architecture has only hydro of 5.89 kW and remaining energy demand is supplied from the grid. The base case system in the summary tab can be changed also there could be scenarios where the base case system is also the winning system.

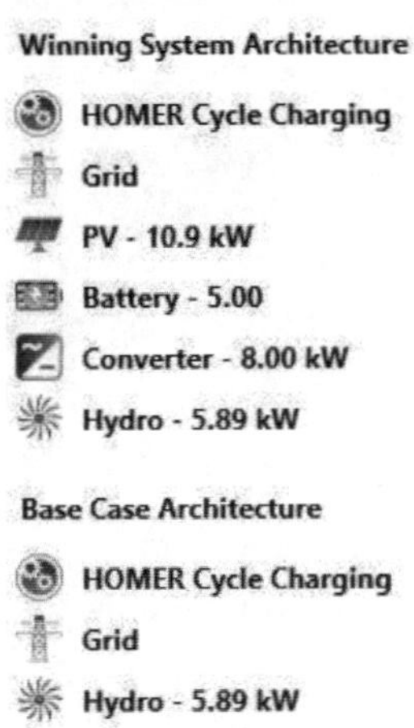

Figure 10. Winning and base case architecture in initial condition

IOP Conf. Series: Materials Science and Engineering 1279 (2023) 012007 doi:10.1088/1757-899X/1279/1/012007

6.4. Winning system architecture

6.4.1. Cost Summary

The Cost Summary tab in the Simulation Results window displays cash flows as either a present value or annualized cost, categorized by component or cost type. Table 9 shows the cash flow based on NPC and categorization is done by the component.

Table 9. Cost Summary of winning system architecture in on-grid system

Component	Capital ($)	Replace ($)	O&M ($)	Fuel ($)	Salvage ($)	Total ($)
Battery	$3,230.50	$6,569.53	$323.19	$0.00	($77.11)	$10,046.11
Generic flat plate PV	$13,703.33	$0.00	$129.28	$0.00	$0.00	$13,832.60
Grid	$0.00	$0.00	$17,408.02	$0.00	$0.00	$17,408.02
Hydro Turbine System	$15,000.00	$0.00	$12,927.52	$0.00	$0.00	$27,927.52
Converter	$961.58	$407.97	$0.00	$0.00	($76.78)	$1,292.76
System	$32,895.41	$6,977.50	$30,788.00	$0.00	($153.89)	$70,507.02

The total capital cost of the system is $32,895 whereas the capital cost of the grid is $0 because the grid supply has been already provided to that community so the cost of extension of the grid is not included. The replacement cost of the hydro turbine and generic flat PV is also $0 because the total lifetime of the project is modeled for 25 years and the lifetime of hydro turbine and PV is also 25 years. Since no diesel generator is added to the system so the total cost of fuel is also $0. The total salvage value at the end of 25 years is $153.59 and the overall cost of the system is $70,507.02.

6.4.2. Cash Flow

There are two ways of displaying the cash flow which is graphical and tabular form. The cash flow in the figure 11 displays the cash flow in graphical form. There are also two options for the cash flow graph which is by cost type and by component type and the figure 11 shows the total inflow or total outflow of cash for a single year based on cost type.

The green bar in 1st year shows the total capital which is $32,895.41. After that $2381.59 is spent every year as an operating cost. Every five years $3230.5 is spent as a replacement cost for the battery and at the end of 25 years, the figure shows a salvage value of $642.41.

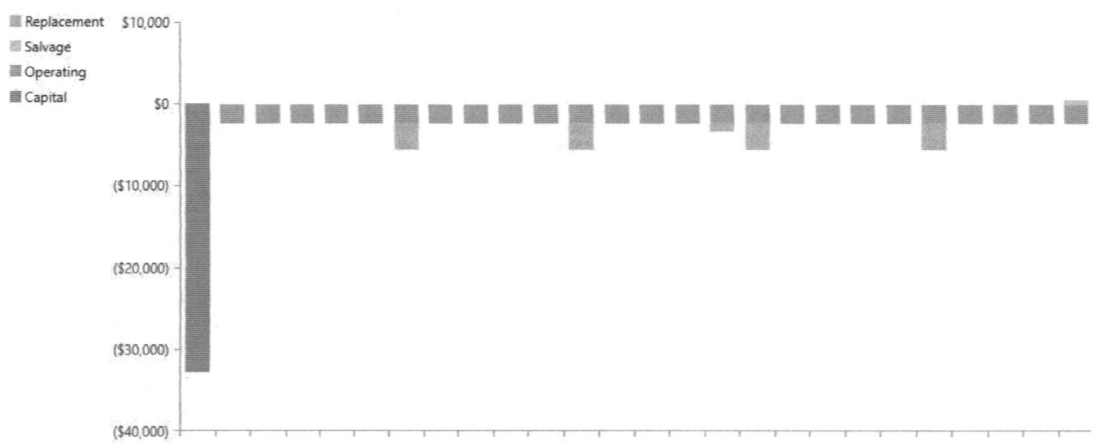

Figure 11. Cash Flow of on-grid system

Based on the components at first year $961.58 is spent for converter, $15,000 for hydro turbine, $13,703.33 for generic flat PV, and 3230.50 for the battery. After that $1346.59 is spent on-grid and $1000 in hydro turbine every year as an operating cost. Every five years, an extra $3255.5 is spent on battery along with other operating costs. At the end of a fifteen-year new converter is also required whose cost is $961.5. If discounted cash flow is kept under consideration, then the operating cost in the first year is $2249.28 and $570.33 on the 25th year at a discount rate of 8%.

6.4.3 Electrical %

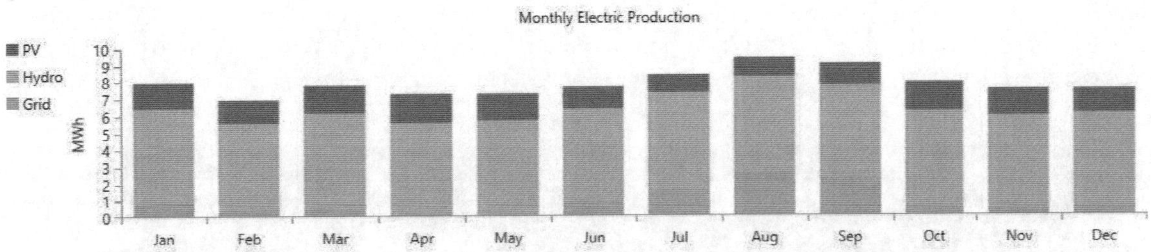

Figure 12. Monthly electric production percentage in on-grid system

The figure 12 shows that out of 100% electric production, Generic flat PV consists of around 18.6% which is 17,609 kWh/yr. Hydro consists of 69.4% which is 65,741 kWh/yr and the rest 12% which is 11,390 kWh/yr is supplied from the grid. Total electricity consumption is 94,740 kWh/yr.

On the consumption part, about 93.3% which is 86,502 kWh/yr is being supplied to AC primary load and the remaining 6.69% is being sold to the grid. This electric percentage is the result of the initial condition. The main factor to vary electric percentage is flow rate. As the flow rate is increased the percentage of hydro in production increases rapidly causing the LCOE of hydro to be very less compared to other sources of energy.

6.4.4. Battery

Figure 13 shows the SOC of the battery for 24 hours. It shows that during the daytime when the sunlight is at its peak the state of charge reaches up to 100% and during morning and night time the state of charge drops down up to 30%.

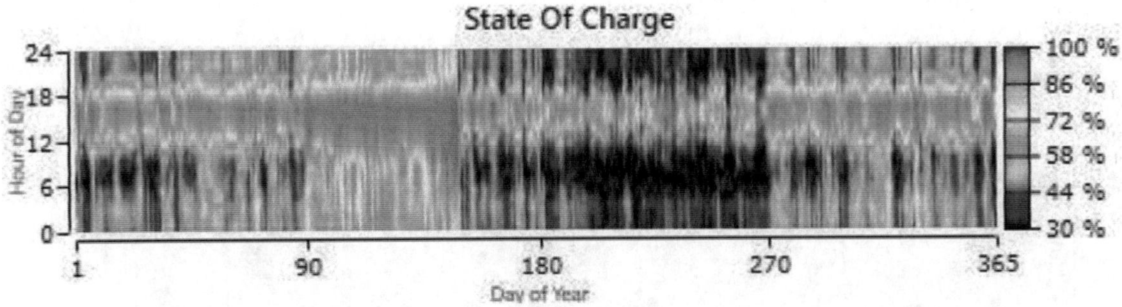

Figure 13. SOC of battery in on-grid system

For this system total number of batteries required is 5. Those batteries will be running at a bus voltage of 12 volts. The autonomy of those batteries is around 5.91 hours which cost around 0.0368 $/kWh. The nominal capacity of the battery is 8.34 kWh whereas the usable nominal capacity is 58.4kWh. The lifetime amount of energy supplied from this storage battery is 89,056 kWh and its expected life is around 5.1 year due to which in every 5-year battery need to be replaced which is clearly explained in the cash flow section.

The total amount of energy into the storage is 17,672 kWh/yr and the total energy out is 17,193 kWh/yr. The total amount of energy depleted in storage is 51.5 kWh/yr and the losses are around 531 kWh/yr. After considering all those parameters, the average energy cost that goes into the storage is around 0.0129 $/kWh.

6.4.5. System converter

Two system converters are used in the system. Inverter for converting DC to AC and rectifier for converting AC to DC. Figure 14 shows power output from the inverter and rectifier for 24 hours throughout the year. The maximum capacity of both inverter and rectifier is 8 kW where the mean output of the inverter is 2.59 kW and rectifier are 0.77 kW with a capacity factor of 32.4% for inverter and 9.63% for the rectifier. The hours of operation of the inverter are 5,499 hrs/yr and the rectifier is 3256 hrs/yr. Total energy out from the inverter is 22,699 kWh/yr and from the rectifier is 6,742 kWh/yr. Total energy into the inverter is 23,862 kWh/yr and into the rectifier is 7,097 kWh/yr. Total losses from the inverter are 1,193 kWh/yr and from the rectifier is 355 kWh/yr.

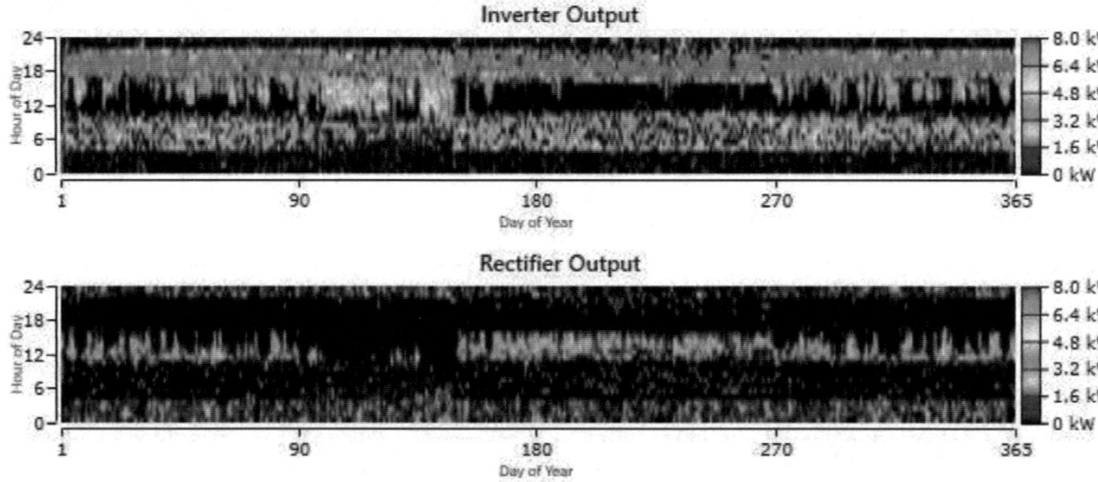

Figure 14. Inverter and rectifier output in on-grid system

6.4.6. PV output

The figure 15 shows PV output for 24 hours throughout a year. From the graph, we can see the output power ranging from 0 kW to 12 kW and output power is at its peak from 12 pm to 2 pm. The total rated capacity of the PV array is 10.9 kW where the mean output is 2.01 kW which is 48.2 kWh/day. The capacity factor is 18.5% and PV penetration is 20.4%. Total production throughout a year is 17,609 kWh/yr in which the hours of operation is 4,381 hrs/yr with an LCOE of 0.0608 $/kWh.

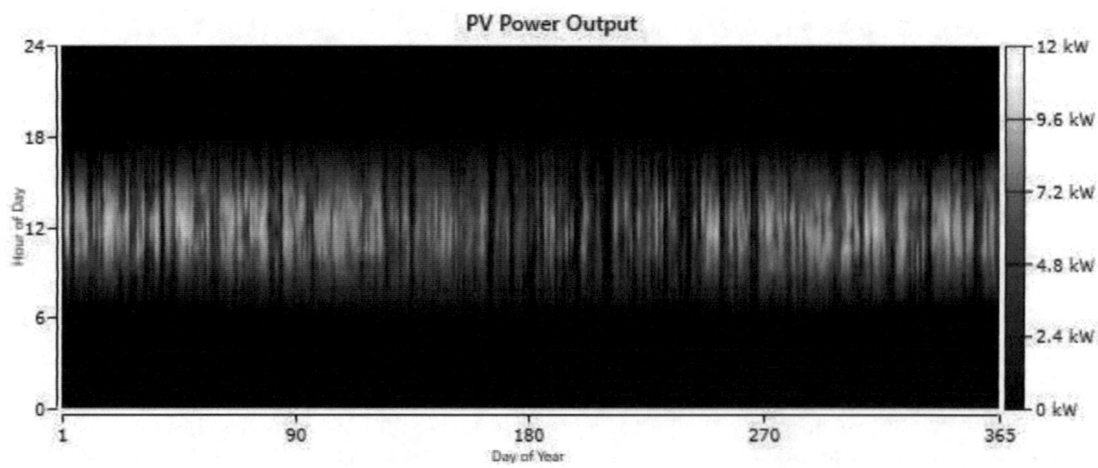

Figure 15. PV output of on-grid system

6.4.7. Grid (Energy Purchased and sold)

The table 10 shows the net energy purchased from the grid along with its peak power. The maximum amount of energy which is 2643 kWh is purchased in August due to the festive season of Nepalese people where there is the highest peak load of 37 kW and the minimum amount of energy which is 47 kWh is sold in May due to the rainy season because the maximum percentage of energy output comes from hydropower.

Table 10. Amount of energy purchased and sold on every month

Month	Energy Purchased (kWh)	Energy Sold(kWh)	Net Energy Purchased (kWh)	Peak Load(kW)	Energy Charge ($)
January	858	499	359	19	$100.13
February	502	432	70	15	$53.03
March	540	556	-15	18	$53.40
April	168	1,368	-1,199	11	($31.12)
May	97	1,544	-1,447	8	($48.23)
June	929	247	682	25	$120.24
July	1,693	47	1,647	26	$235.19
August	2,643	50	2,594	37	$368.09
September	2,353	40	2,313	37	$327.84
October	632	378	254	17	$73.39
November	497	608	-110	20	$45.29
December	476	433	43	17	$49.34
Annual	11,390	6,200	5,190	37	$1,346.59

7. Results for Off-Grid System

7.1. Sensitivity Variables

Table 11 shows the lists of sensitivity variables that are used during this simulation. Two sensitivity variables were taken which were Hydro capital cost and Hydro flow rate.

Initial condition:
Hydro Capital Cost – 15000 $ Hydro Flow rate – 75 L/S

Table 11. List of sensitivity variables in an off-grid system

Hydropower Cost($/kWh)	Hydro flow rate(L/s)
15000	50
18000	75
20000	100
22000	
25000	

The table 12 shows the results of the base case and lowest case system while changing the Hydropower cost at the initial condition. From the table 12, we can conclude that when hydropower cost is increased then NPC and LCOE increase but operating cost remains constant.

The table 13 shows the results of the base case and lowest case system while changing the Hydro flow rate at the initial condition. From the table 13, we can conclude that when the hydro flow rate is decreased then NPC, LCOE and operating cost increase.

Table 12. Sensitivity analysis when hydropower cost is varied in off-grid system

Hydropower cost(L/s)	Parameters	System
15000	NPC	$108,210
	Operating Cost	$1749/yr
	LCOE	$0.09683/kWh
18000	NPC	$111,210
	Operating Cost	$1,749/yr
	LCOE	$0.09951/kWh
20000	NPC	$113,210
	Operating Cost	$1749/yr
	LCOE	$0.1013/kWh
22000	NPC	$115,210
	Operating Cost	$1,749/yr
	LCOE	$0.1031/kWh
25000	NPC	$118,210
	Operating Cost	$1749/yr
	LCOE	$0.1058/kWh

The table 13 shows the results of the base case and lowest case system while changing the Hydro flow rate at the initial condition. From the table 13, we can conclude that when the hydro flow rate is decreased then NPC, LCOE and operating cost increase.

Table 13. Sensitivity analysis when hydro flow rate is varied in an off-grid system

Hydro flow rate(L/s)	Parameters	System
100	NPC	$72,520
	Operating Cost	$1,483/yr
	LCOE	$0.06489/kWh
75	NPC	$108,210
	Operating Cost	$1749.57/yr
	LCOE	$0.09683/kWh
50	NPC	$149,885
	Operating Cost	$2,247/yr
	LCOE	$0.1341/kWh

7.2. List of all the system architecture

The results of HOMER Pro simulation and optimization processes showed that among the 78,260 system solutions, only 28,900 were feasible. 49,360 solutions were infeasible due to capacity shortage constraints. 19,850 were omitted and among them, 10,550 were lacking convertor, 3,810 were having unnecessary converter and 5,070 were for no sources of power generation.

Architecture						Cost			
PV (kW)	G10	PowerSafe SBS 1200	Hyd100 (kW)	Converter (kW)	Dispatch	NPC ($)	COE ($)	Operating cost ($/yr)	Initial capital ($)
38.1		28	5.89	36.8	CC	$108,211	$0.0968	$1,750	$85,593
37.5	1	27	5.89	44.5	CC	$187,076	$0.167	$2,941	$149,055
99.8		81		86.9	CC	$215,628	$0.193	$2,087	$188,648
190	1	49		42.9	CC	$371,254	$0.332	$2,425	$339,910
	36	224	5.89	148	CC	$3.12M	$2.79	$49,247	$2.48M

Figure 16. List of all system architecture in off-grid system

IOP Conf. Series: Materials Science and Engineering 1279 (2023) 012007 doi:10.1088/1757-899X/1279/1/012007

A total of 5 system architectures were proposed by HOMER Pro. Based on the cost of electricity and operating cost, the system architecture for the lowest NPC which is $0.0649 is the system having 16 batteries, 36.8 kW of the converter, 18.7 kW of PV and 7.85 kW of Hydro. The figure above also shows that the system with turbines has the highest LCOE, NPC and operating cost.

7.3. Winning system and base system architecture

In this system architecture, only one architecture is shown because the base case architecture system wins so winning system architecture is also the same in this case. At initial condition, the system architecture has hydro of 3.92 kW, converter of 39.5 kW and PV of 54.6 kW along with the grid. Whenever any sensitivity variables are changed then system architecture also changes as it was with the grid system.

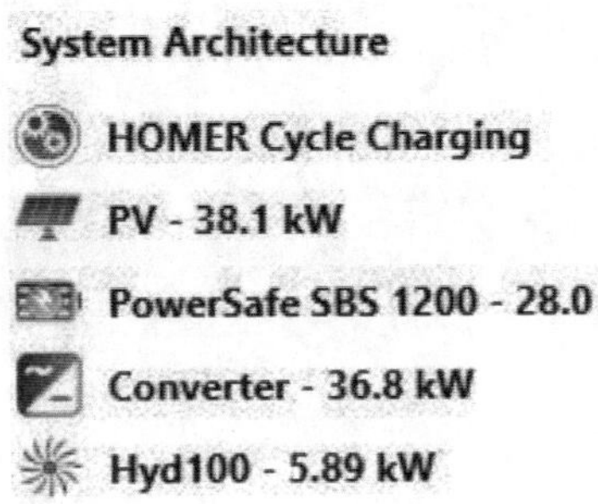

Figure 17. System architecture in initial condition

7.4. Winning system architecture

7.4.1. Cost Summary

The table 14 shows the cash flow on a basis of NPC and categorization is done by the component.

Table 14. Cost summary in off-grid system

Component	Capital ($)	Replace ($)	O&M ($)	Fuel ($)	Salvage ($)	Total ($)
Battery	$18,090.80	$7,675.75	$1,809.85	$0.00	($1,444.60)	$26,131.51
Generic flat plate PV	$48,089.06	$0.00	$129.28	$0.00	$0.00	$48,218.34
Hydro Turbine	$15,000.00	$0.00	$12,927.52	$0.00	$0.00	$27,927.52
System Converter	$4,413.37	$1,872.48	$0.00	$0.00	($352.42)	$5,933.43
System	$85,593.23	$9,547.93	$14,866.64	$0.00	($1,797.02)	$108,210.79

The total capital cost of the system is $85,593. The total replacement cost of the system is $9,547.93 whereas hydro turbine and generic flat PV is $0 because the total life time of the project is modeled for 25 years and the lifetime of hydro turbine and PV is also 25 years. The total Operating and Maintenance cost is $14,866.64. Since no diesel generator is added to the system so the total cost of fuel is also $0. The total salvage value at the end of 25 years is $1797.02 and the overall cost of the system is $108,210.79.

7.4.2. Cash Flow

Figure 18 displays the cash flow in graphical form based on component type.

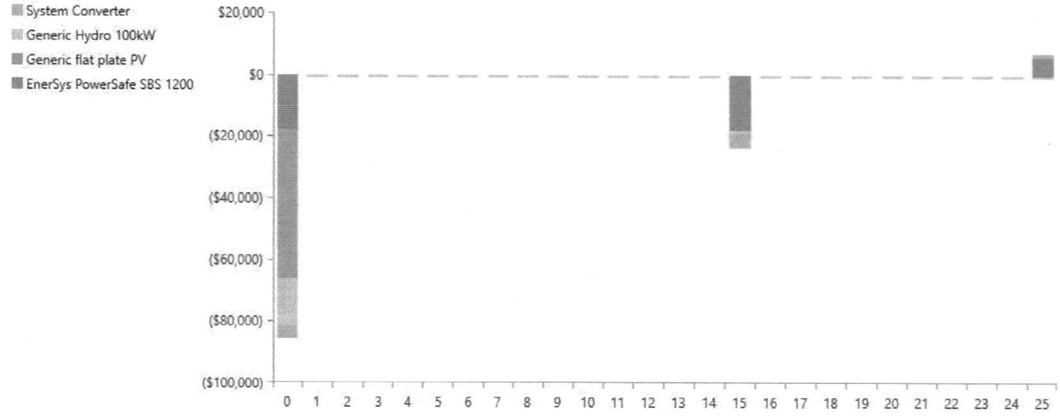

Figure 18. Cash flow in off-grid system

The bar in 1st year shows the total capital which is $85,593.23. Every year $1000 is spent for the operation of hydropower nominally. If operation cost is also discounted, then its value decreases year by year. After 15 years $18230.80 is spent on replacing the battery and $4413.47 is spent on the converter. These two prices are the nominal cash flow but their total discounted cash flow will be $9547.93.

Based on a cost type, the total capital spending in the first year is the same. After that $2381.59 is spent every year as an operating cost. On every fifteen years, $22,504.17 is spent as a replacement cost and at the end of 25 years, the graph shows a salvage value of $7,501.39.

7.4.3. Electrical %

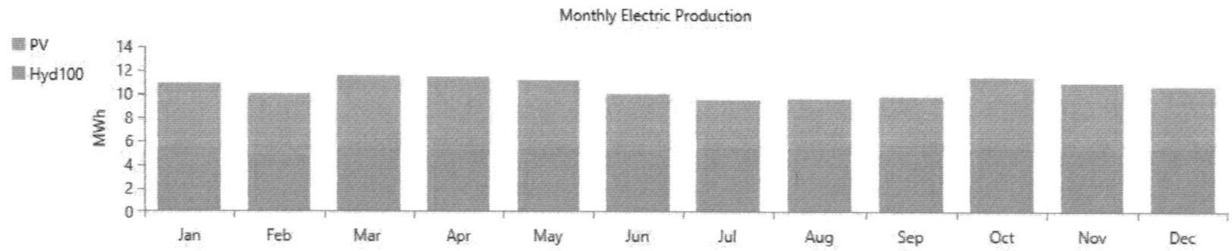

Figure 19. Monthly electric production percentage in an off-grid system

Figure 19 shows that out of 100% electric production, Generic flat PV consists of 48.4% which is 61,703 kWh/yr and Hydro consists of 51.6% which is 65,741 kWh/yr. Total electricity production is 127,444 kWh/yr and consumption is 86,447 kWh/yr. On the consumption part, about 100% energy is being supplied to AC primary load. Since the rate of energy production is higher than the rate of energy consumption in this system about 30.3% which is 38,586 kWh/yr amount of energy will be wasted. The unmet electric load is 53.9 kWh/yr which is nearly negligible.

7.4.4. Battery

Figure 20 shows the SOC of the battery for 24 hours. It shows that from January to July the SOC is above 72% and during day time it's around 100 %. But in August and September SOC is in the range of 30 % to 50% and also during day time it's not reaching to 100%. But after September SOC is the same as its initial months.

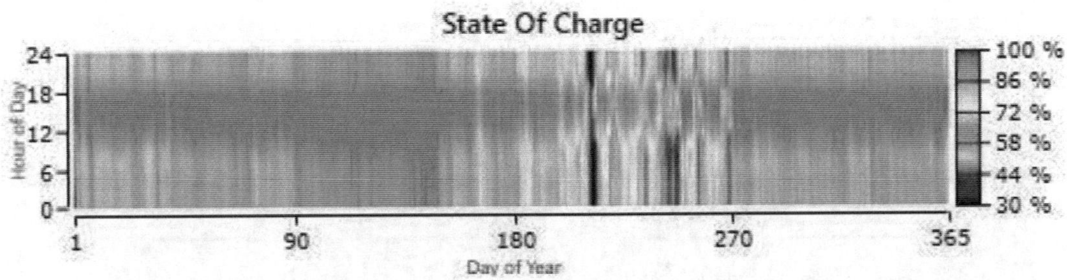

Figure 20. SOC of battery in off-grid system

For this system total number of batteries required is 28. Those batteries will be running at a bus voltage of 12 volts. The autonomy of those batteries is around 33.1 hours which cost around 0.0368 $/kWh. The nominal capacity of the battery is 467 kWh whereas the usable nominal capacity is 327 kWh. The lifetime amount of energy supplied from this storage battery is 368,516 kWh and its expected life is around 15 years due to which in every 5-year battery needs to be replaced which is clearly explained in the cash flow section.

The total amount of energy into the storage is 24,890 kWh/yr and the total energy out is 24,19 kWh/yr. The total amount of energy depleted in storage is 54.2 kWh/yr and the losses are around 748 kWh/yr.

7.4.5. Inverter and rectifier output

Figure 21 shows power output from the inverter and rectifier for 24 hours throughout the year. The maximum capacity of both inverter and rectifier is 36.8 kW where the mean output of the inverter is 3.34 kW and the rectifier is 0.389 kW with a capacity factor of 34% for inverter and 9.09% for the rectifier. The hours of operation of the inverter are 4,852 hrs/yr and the rectifier is 2,217 hrs/yr. Total energy out from the inverter is 29,264 kWh/yr and from the rectifier is 3,408 kWh/yr. Total energy into the inverter is 30,804 kWh/yr and into the rectifier is 3,587 kWh/yr. Total losses from the inverter are 1,540 kWh/yr and from the rectifier is 179 kWh/yr.

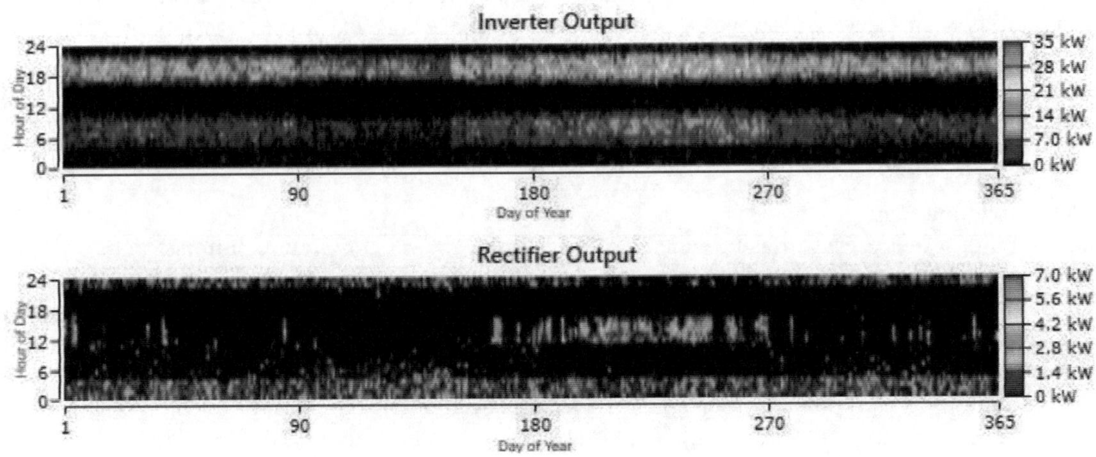

Figure 21. Inverter and rectifier output in off-grid system

7.4.6. PV output

Figure 22 shows PV output for 24 hours throughout a year. The total rated capacity of the PV array is 38.1 kW where the mean output is 7.04 kW and the maximum output is 37.7 kW which is 169 kWh/day. The capacity factor is 18.5% and PV penetration is 71.3%. Total production throughout a year is 61,8703 kWh/yr in which the hours of operation is 4,381 hrs/yr with an LCOE of 0.0604.

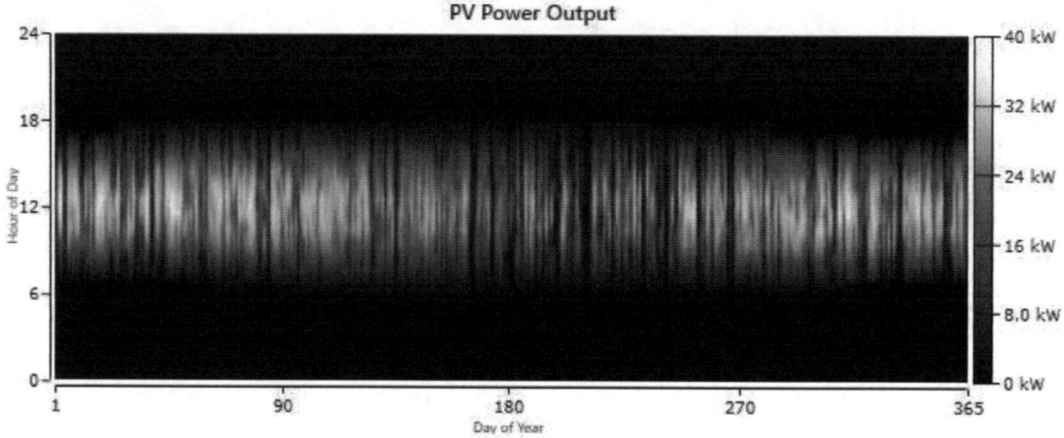

Figure 22. PV output in off-grid system

8. Conclusion

This paper presents a comparative analysis between an off-grid and a grid-connected hybrid power system. The site selected for this project is Thakle Namuna Basti a community village built after the earthquake of 2015. The resources required for this project were provided by NREL and Meteonorm. Two types of results are presented in this paper. Sensitivity analysis was carried out with hydropower cost, the flow rate of the river, sellback and the purchase price of electricity which showed that for the lowest NPC and LCOE the sellback rate of electricity and the flow rate of the river should be high and purchased power price and hydropower cost should be as low as possible. Optimization results showed that for the on-grid system, the system with hydro and grid is the most feasible for the community based on NPC and LCOE. For off-grid systems, the system with PV, Hydro, and Battery is the most feasible to meet the demand of the community village. The price of electricity is higher for an off-grid system compared to an on-grid system but the simulation of an off-grid system was also done because there is a continuous increase in consumption demand which is exerting extra pressure on the central grid. Besides this, any issue in the system may result in a complete shutdown when the whole system is dependent on only the central grid. But again, in an off-grid system, extra battery banks are required for energy storage but in an on-grid system, an extra amount of electricity can be sold to the grid which will also work as a source of revenue.

9. Acknowledgement

The funding for the research was provided jointly through University Grant Commission (UGC) Nepal and Energize Nepal Program, Kathmandu University. Also, we are eternally grateful to Mr. Khil Bahadur Tamang for his assistance during the survey and field visits.

10. References

[1] Rajbongshi R, Borgohain D, Mahapatra S. Optimization of PV-biomass-diesel and grid base hybrid energy systems for rural electrification by using homer. Energy. 2017;126:461–74.
[2] Saheb-Koussa D, Haddadi M, Belhamel M. Economic and Technical Study of a hybrid system (wind– photovoltaic–diesel) for Rural Electrification in Algeria. Applied Energy. 2009;86(7-8):1024–30.
[3] Azimoh CL, Klintenberg P, Mbohwa C, Wallin F. Replicability and scalability of mini-grid solution to rural electrification programs in sub-saharan africa. Renewable Energy. 2017;106:222–31.
[4] Murugaperumal K, Ajay D Vimal Raj P. Feasibility design and techno-economic analysis of hybrid renewable energy system for rural electrification. Solar Energy. 2019;188:1068–83.

[5] Baneshi M, Hadianfard F. Techno-economic feasibility of hybrid diesel/PV/wind/battery electricity generation systems for non-residential large electricity consumers under southern Iran climate conditions. Energy Conversion and Management. 2016;127:233–44.

[6] Iea. Electricity Information - Data product [Internet]. IEA. [cited 2022Nov3]. Available from: https://www.iea.org/data-and-statistics/data-product/electricity-information

[7] Nepal Electricity Authority. A year in review-fiscal year 2019/2020 p. 8–13.

[8] Gunatilake H, Wijayatunga P, Holst DR-. HYDROPOWER DEVELOPMENT AND ECONOMIC GROWTH IN NEPAL. ASIAN DEVELOPMENT BANK; 2020.

[9] Kanase-Patil AB, Saini RP, Sharma MP. Integrated Renewable Energy Systems for off grid rural electrification of remote area. Renewable Energy. 2010;35(6):1342–9.

[10] Li C, Ge X, Zheng Y, Xu C, Ren Y, Song C, et al. Techno-economic feasibility study of autonomous hybrid wind/PV/battery power system for a household in Urumqi, China. Energy. 2013;55:263–72.

[11] Abnavi MD, Mohammadshafie N, Rosen MA, Dabbaghian A, Fazelpour F. Techno-economic feasibility analysis of stand-alone hybrid wind/photovoltaic/diesel/battery system for the electrification of remote rural areas: Case study Persian Gulf coast-iran. Environmental Progress & Sustainable Energy. 2019;38(5):13172.

[12] Google maps. Google; [cited 2022 Nov3]. Available from: https://www.google.com/maps/@27.805681,85.547194,173m/data=!3m1!1e3

[13] Bunnoon P, Chalermyanont K, Limsakul C. Mid-term load forecasting: Level suitably of wavelet and neural network based on factor selection. Energy Procedia. 2012;14:438–44.

[14] Wang C-hsiang, Grozev G, Seo S. Decomposition and statistical analysis for regional electricity demand forecasting. Energy. 2012;41(1):313–25.

[15] NASA Power [Internet]. NASA. NASA; [cited 2022Nov3]. Available from: https://power.larc.nasa.gov/

[16] Streamflow prediction tool - nepal. [cited 2022Nov3]. Available from: http://tethys.icimod.org/apps/streamflownepal/

[17] Homer pro 3.15 [Internet]. Welcome to HOMER. [cited 2022Nov3]. Available from: https://www.homerenergy.com/products/pro/docs/latest/index.html

[18] NewsTeam. Solar Panel Price in Nepal [Internet]. ktm2day.com. 2022 [cited 2022Nov3]. Available from: https://www.ktm2day.com/solar-panel-price-nepal/

[19] Chung I-Y, Liu W, Cartes DA, Schoder K. Control parameter optimization for a microgrid system using particle swarm optimization. 2008 IEEE International Conference on Sustainable Energy Technologies. 2008;

Production of fuels by pyrolysis of waste plastics: technical notes

Saroj Raj Kafle[1*], Pragati Gaire[1], Nerisha Tuladhar[1]

[1] Department of Chemical Science and Engineering, Kathmandu University, PO BOX 6250, Dhulikhel, Kavre, Nepal
*Corresponding autho: sarojrajkafle@gmail.com

Abstract. This paper described the plant scale catalytic process of the plastic waste materials in order to obtain an alternative fuel from three plastic wastes Polypropylene (PP), Polystyrene (PS) and Polyethylene (PE). Pyrolysis is the thermochemical decomposition of plastics at elevated temperatures (in absence or little supply of oxygen) into a range of useful products: gas, liquid, and solid residue. These waste have plastics consisting of PP, PE and PS, are first sorted manually and crushed by using crusher and finally feed in the reactor enclosed in the burner maintaining the reaction temperature about 500°C. Nitrogen gas is used to create inert environment and for fluidization in reactor. The produced cracked gas is now sent to be condensed to a condenser followed by the series of cyclone separator and electrostatic precipitator then finally feed into the liquid vapor separator. The plastic pyrolysis plant is design of capacity 3.2 tons of plastics each day, obtained 80% gaseous fuel and 7% Liquid and remaining as residue. The variation of yield of the products according to increase in temperature is also shown in this paper.

Keywords: Plastic Waste, Pyrolysis, Reactor, Crack gas

1. Introduction

For more than 50 years, plastic has played an important part in improving people's lives. Plastic can be found everywhere, from the summit of Mount Everest to the depths of the ocean. Mount Everest holds approximately 100,000 kilograms of plastic, and the ocean contains approximately 1.5 billion tons of plastic. The amount of plastic waste generated each year is expected to rise at a rate of 3.9% per year [1]. This, along with the existing volume of municipal solid trash, makes plastic waste management an ever-growing issue. Furthermore, as many countries seek greater energy independence, there is a growing interest in alternative fuels, with waste-derived feedstocks such as waste biomass and plastic waste gaining the most attention due to their abundance and environmental benefits.

Due to their low price and good qualities, polyethylene (PE), polypropylene (PP) and polystyrene (PS) are regularly used plastics than Polyvinyl Chloride (PVC). PVC contain chlorinated compound which are hazardous catalyst poisonings and have acidic character that can deteriorate the fuel quality. It also produced the major product of harmful hydrochloric acid and very low yield of liquid oil. In case of PS, it can degrade easily without the need of any catalysts to speed up the reaction. Therefore, PS was considered as the best plastic for pyrolysis because it produced the most liquid oil of all the plastics tested [2]. Polyethylene (PE) is the most commonly used plastic in Europe with 29.8%, followed by polypropylene with 19.3% and polystyrene with 6.7%.

Among various techniques, thermochemical conversion technologies such as pyrolysis, incineration and gasification, have gained significant attention for the management of plastic waste.

Pyrolysis, in particular, has demonstrated substantial advantages over the others since it produces fewer gaseous pollutants due to the lack of oxygen in the process. It is the process of using heat and pressure to break down long chain polymer molecules into smaller, less complex molecules. The process necessitates intense heat for a short period of time in the absence of oxygen. The pyrolysis process is promising because it produces wax/oil as a heavy fuel alternative or as a raw material for resale into the petrochemical industry, as well as char and gases that may be sold or further processed to add value to the overall process. Many researchers chose pyrolysis because it produces a large amount of liquid oil, up to 80 wt. %, at a moderate temperature of 500°C [4] .The liquid oil generated can be used in a variety of applications, including furnaces, boilers, turbines, and diesel engines, without the need for further treatment or upgrading. It does not cause water contamination and is considered as green technology when even the pyrolysis byproduct which is gaseous has substantial calorific value that it can be reused to compensate the overall energy requirement of the pyrolysis plant [5]. The Yield of different combination of plastic is shown in Table 1. According to [6], PE, PP and PS are regarded as the best plastic combination for the pyrolysis process. So in this paper we chose the PE, PP And PS mixture as a suitable plastic. The primary objective of this project is to design a pyrolysis plant based on the amount of wastes generated for the city of Kathmandu. The plant aims to produce about 2.5 tons of fuel gas, 0.224 tons of fuel oil per day from an estimated 3.2 tons waste per day.

Table1. Yield of different combination of plastics

Feed	Pyrolysis Type	Gas Yield (%)	Liquid Yield (%)	Char Yield (%)	Reference
PP	Thermal degradation	17	70	13	[7]
PE	Thermal degradation	14	70	16	[8]
PE,PP,PET	Thermal degradation	16	80	4	[8]
PET,PS	Thermal degradation	15	63	22	[9]
PE,PP,PS	Thermal degradation	46.6	52	1.4	[10]
PS	Thermal degradation	9.9	24.6	65.5	[11]
All	Catalytic degradation	56.9	40.4	3.2	[12]
PP+PE+PS	FCC	80.3	7.4	12.3	[12]

2. Materials and Method

The general process flow diagram of the process is shown in the Figure 1 and overall plant flow diagram is shown in Figure 2. The raw material consists of only PP, PE and PS plastics. These plastics are manually sorted as much as possible. The sorted plastics are then passed to the Crusher (CR-101) for size reduction. The product of crusher is then passed through a screen(S-101) which lets the plastics pieces which are under the size of 6 mm pass through it and the oversize product is recycled to the crusher. The products are then transported to the reactor (R-101) by series of Fuller Kinyon pump (P-101) and Rotary Dryer (DR-101) in which an inert environment is created by passing nitrogen gas through it at the rate of 200 ml/min. The reactor is enclosed within the burner which provides the necessary energy as heat for the cracking of plastics into alkane, alkene, aromatics and other simpler products. The burner is a husk fired burner (H-101). Rice husk is used as the fuel source to fire up the burner. A combination of 25% excess air and rice husk at the flow rate of about 750 kg/day is enough to fire up the reactor. The reaction temperature is about 500°C. So, husk fired burner provides the heat. The catalyst used is equilibrium catalyst also known as zeolite catalyst. Zeolite is a compound formed by the mixture of sodium, aluminum, silicon, oxygen and hydrogen. Average chemical composition of sodium zeolite is reported as Sodium Oxide – 17%, Aluminum Oxide – 28%, Silicon dioxide – 33% and water – 22%. Molecular formula of zeolite can be generally represented as $NaAlSi_2O_6$-H_2 [10]. The amount of catalyst used per day is about 350 kg. This amount is feed to the reactor before the system runs and by the end of the day, the catalyst is nearly dead. The catalyst can be regenerated and reused adding small amount of fresh catalyst next day. The residence time of the plastic is about half an hour and the produced gas has residence time of about two seconds. The reactor should be completely airtight and fluidization is maintained by purging compressed nitrogen gas at the rate of 600 ml/min. The reaction mostly occurs at the minimum fluidization. The produced cracked gas is passed through the series of Cyclone Separator(C-101) and Electrostatic Separator (E-101) for separation from solids particles and finally sent to a condenser (H-102). The produced cracked product from gas liquid separator (F-101) consists of about 80% fuel gas, and 7% fuel oil and remailing as solid char.

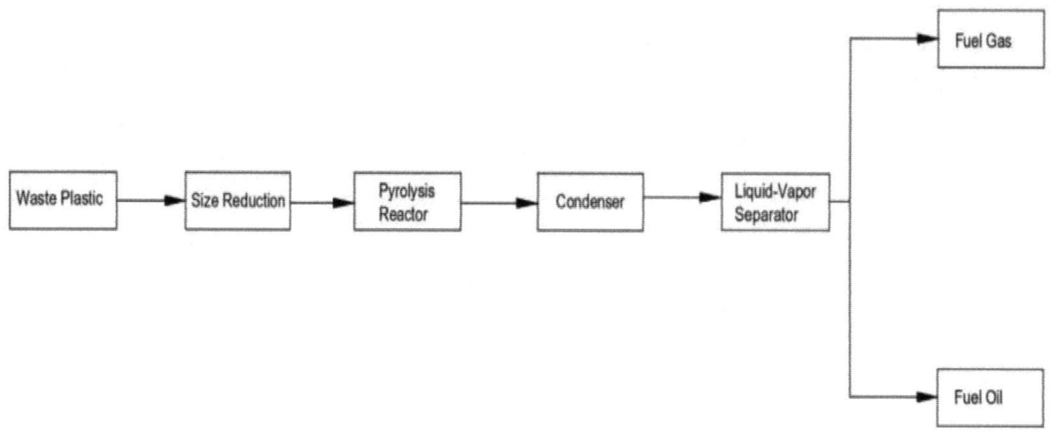

Figure 1 Basic flow diagram

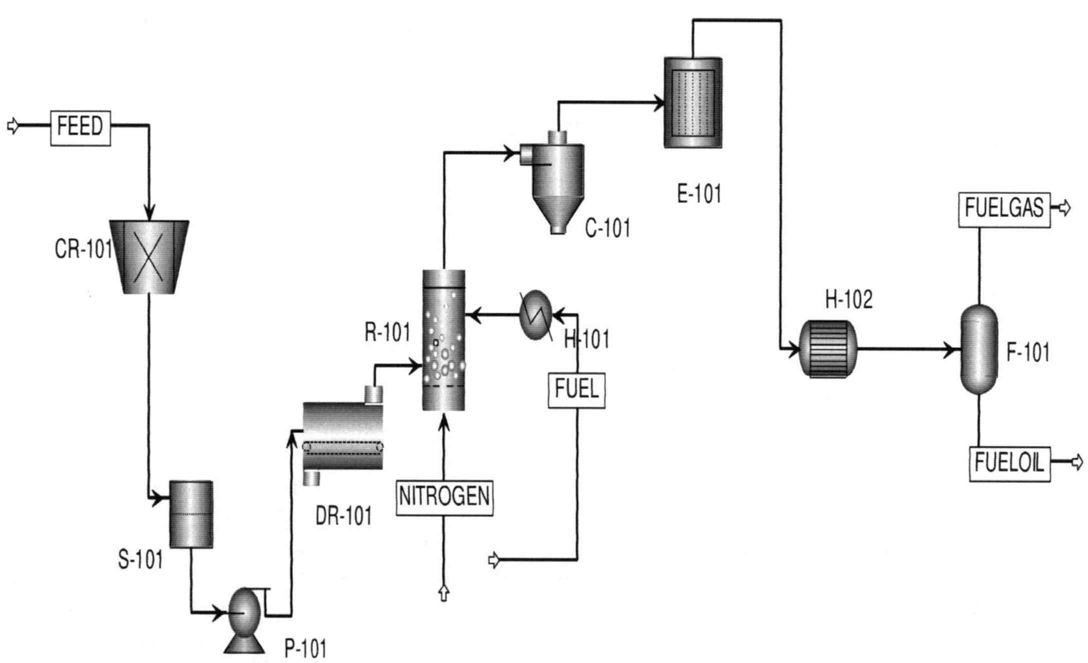

Figure 2. Overall plant flow diagram

3. Results and Discussion

The produced cracked product consists of about 80% fuel gas, and 7% fuel oil and remaining as solid char. The solid char can also be used another field like road construction. The Fourier Transform Infrared Spectroscopy analysis of producing fuel oil and fuel gas obtained by pyrolysis having more than 30 compounds in significant quantity which is shown in the Table 2 [13].

Table 2. Qualitative Analysis of Product

Type of Material	Component	Type of Material	Component
	Methane		n-nonane
	Ethane		n-decane
	Propane	**Aromatics**	n-undecane
C1-C4	Butane		n-undecane
	Ethene		n-dodecane
	Propene		1-pentene
	Butene		1-hexene
	Benzene		1-heptene
	Ethylbenzene	**Aliphatic/ Alkane**	1-octene
	o-xylene		1-nonene
	p-xylene		1-decene
	Toluene		1-undecene
Aromatics	n-propyl benzene		1-dodecene
	n-butyl benzene		cycloheptane
	n-pentane		cyclooctane
	n-hexane	**Cycloalkane**	cyclohexane
	n-heptane		cyclopentane
	n-octane		cyclobutane

The Feed and Product Specification is Calculated by Performing Mass and Energy Balance which is shown in the Table 3, Table 4 and Table 5. With respect to the solid fraction(residue), PE, PP and PS the product composition varies. So, with the feed specification shown in Table 3, the product compositions are taken which is shown in Table 4. The Quantitative Gas Chromatography analysis of products obtained after the pyrolysis are shown in the Table 5. The Heat of formation value is taken from the reference [15].

Table 3. Feed Specification [9]

Feed	Weight (%)	Flow Rate (Kg/h)	Density (Kg/m^3)	Flow Rate (m^3/h)
PE	31	62	910	0.068
PP	62	124	935	0.132
PS	7	14	1060	0.0132
Catalyst	-	333.33 (kg/day)	2400	0.138
N_2	-	800 (ml/min)		

Table 4. Product Specification [9]

Products	Compositions (%)	Flowrate (Kg/h)
Gas	80.3	160.6
Liquid	7.4	14.8
Residue	12.3	24.6

IOP Conf. Series: Materials Science and Engineering 1279 (2023) 012008 doi:10.1088/1757-899X/1279/1/012008

Table 5. Heat of Formation of Products [15]

Type of Material	Component	Mass Flow Rate (kg/h)	Formation Energy of the Material at 400°C(kJ/mol)
C1-C4	Methane	1.23	−1.21
	Ethane	2.14	−1.09
	Propane	10.01	−3.90
	Butane	4.79	−1.58
	Ethene	0.65	0.49
	Propene	1.52	0.57
	Butene	1.30	0.30
Aromatics	Benzene	10.07	4.77
	Ethylbenzene	31.80	8.78
	O - xylene	5.83	1.32
	M - xylene	2.91	0.70
	P - xylene	2.91	0.70
	Toluene	11.64	3.95
	N - Propylbenzene	5.83	1.30
	N - Butylbenzene	8.73	1.63
Aliphatic/Alkane	N - pentane	3.46	−1.04
	N - hexane	3.46	−0.96
	N - heptane	3.46	−0.90
	N - octane	3.46	−0.86
	N - nonane	3.46	−0.83
	N - decan	3.46	−0.79
	N - undecan	3.46	−1.20
	N - dodecan	3.46	−1.016
Aliphatic/Alkene	1 - pentene	3.46	0.54
	1 - hexene	3.46	0.36
	1 - heptene	3.46	0.23
	1 - octene	3.46	0.15
	1 - nonene	3.46	0.06
	1 - decene	3.46	0.011
	1 - undecene	3.46	0.08
	1 - dodecene	3.46	−0.09
Cycloalkane	Cycloheptane	3.69	−0.41
	Cyclooctane	3.69	−0.32
	Cyclohexane	3.69	−0.63
	Cyclopentane	3.69	−0.26
	Cyclobutane	3.69	1.37
Solid Residue		24.6	2.17
Total		175.400	12.40

Heat of formation of reactants and products are calculated from energy balance and from that value the amount of energy that should be continuously supplied to the reactor to carry out the complete cracking is calculated as 114.68 KW.

According to [14], the yield of the liquid products increases with increase in pyrolysis temperature whereas decreases in case of gaseous product. The yield distribution of the product obtained from pyrolysis is shown in the Figure 3.

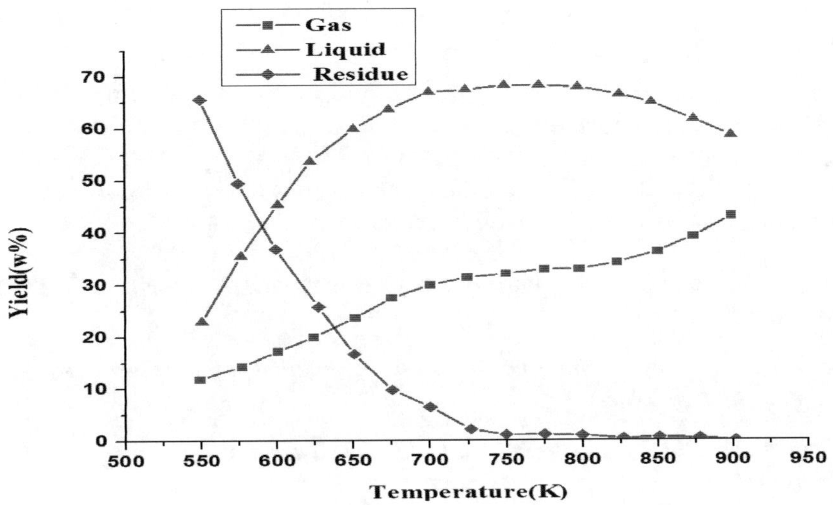

Figure 3. Variation of Temperature on Yield of Products

4. Conclusions

The plastic pyrolysis plant is design of capacity 3.2 tons of plastics each day, obtained 80% gaseous fuel and 7% Liquid and remaining as residue. This process used a mixture of LDPE (Low Density Poly Ethylene), HDPE (High Density Poly Ethene), PP (Poly Propylene) and PS (Poly Styrene) to produce fuel oil and fuel gas. A fluidized bed catalytic reactor for plastic pyrolysis has been designed. The system also comprises of condenser, gas – vapor separator, and burner as major equipment. This reactor designed is more compact and smaller in size compared to the existing reactors. Process and design parameters can be varied to give product ratio in desired proportion. Reaction Temperature plays vital role in the yield of the products. The produced fuel oil can also be further processed to produce synthetic fuel oil. The synthetic fuel oil has lower carbon emission ratio, this can on one hand manage the waste plastic problem to some extent and on other hand can also reduce global warming process.

5. REFERENCE

[1] Europe, Plastic 2015. *An analysis of European plastics production, demand and waste data*

[2] Sharuddin SD, Abnisa F, Daud WM, and Aroua MK 2018. *Pyrolysis of plastic waste for liquid fuel production as prospective energy resource*

[3] FakhrHoseini, Seyed Mousa, and Majid Dastanian 2013. *Predicting pyrolysis products of PE, PP, and PET using NRTL activity coefficient model*

[4] Abnisa, Faisal, and Wan Mohd Ashri Wan Daud 2014. *A review on co-pyrolysis of biomass: an optional technique to obtain a high-grade pyrolysis oil*

[5] Kunwar B, Cheng HN, Chandrashekaran SR, and Sharma BK 2016. *Plastics to fuel: a review*

[6] Saputra GA, Purnomo CW, Prawisudha P, and Sulistyo H 2019. *Kinetics Modeling of Waste Plastic Mixture Pyrolysis for Liquid Fuel Production*

[7] Li, Haoyu. 2017. *Applications of lumping kinetics methodology to plastic waste recovery via pyrolysis, Heriot-Watt University*

[8] Al-Salem, S. M., and P. Lettieri 2010. *Kinetics of polyethylene terephthalate (PET) and polystyrene (PS) dynamic pyrolysis*

[9] Buekens, A.G. and Huang, H 1998. *Catalytic plastics cracking for recovery of gasoline-range hydrocarbons from municipal plastic wastes*

[10] Miandad, R., M. A. Barakat, Asad S. Aburiazaiza, M. Rehan, and A. S. Nizami 2016. *Catalytic pyrolysis of plastic waste: A review*

[11] Lin, Y.H. and Yang, M.H 2005. *Catalytic reactions of post-consumer polymer waste over fluidised cracking catalysts for producing hydrocarbons*

[12] Miandad, R., M. A. Barakat, Asad S. Aburiazaiza, M. Rehan, and A. S. Nizami 2016. *Catalytic pyrolysis of plastic waste*

[13] Dash, Abhishek 2012. *Study on the thermal pyrolysis of medical waste (plastic syringe) for the production of useful liquid fuels*

[14] Demirbas, Ayhan 2004. *Pyrolysis of municipal plastic wastes for recovery of gasoline-range hydrocarbons*

[15] Smith, Joseph Mauk 2016. *Introduction to Chemical Engineering Thermodynamics. McGraw Hill Education.*

IOP Conf. Series: Materials Science and Engineering 1279 (2023) 012009 doi:10.1088/1757-899X/1279/1/012009

Numerical Investigation of Flow Behaviour and Erosion Potential in the Side-Wall Clearance Between Guide Vane and Runner

A Pandey[1*], S Gautam[2], S Chitrakar[1], N Acharya[3], P L Bijukchhe[1]
[1]Department of Mechanical Engineering, Kathmandu University, Dhulikhel, Nepal
[2]Turbine Testing Laboratory, Kathmandu University, Dhulikhel, Nepal
[3]Waterpower Laboratory, Norwegian University of Science and Technology, Trondheim, Norway
[*]Corresponding Author: arun.pandey@ku.edu.np

Abstract. In the hydropower plants exposed to excessive sediment concentration, the components of Francis turbines, especially the regions of Guide Vanes (GVs) and Runner Vanes (RVs) are severely eroded. The side-wall clearance between GV and runner is present in these turbines to separate the rotating and stationary components. The height difference between the runner inlet and GV outlet covers, introduced during the fabrication or assembly process might play a crucial role in accelerating the erosion process in these components. This study is focused on numerical analysis of simplified geometries that replicate different scenarios of the height difference and consequent flow behaviour in the side-wall clearance of Francis turbines. The numerical analysis is done in OpenFOAM using the available solver adopting the finite volume method (FVM). The numerical study is compared with the experimental results obtained from Rotating Disc Apparatus (RDA). The numerical analysis shows the vortex formation in the side-wall clearance region which traps the sediment particles. Such vortex formation causes erosion in the region. The vortex formation and flow behaviour in different height difference scenarios vary with each other and so does the erosion pattern in the experimental results.

Keywords: Sediment erosion, Francis turbine, Side-wall clearance, OpenFOAM, CFD

1. Introduction

The sediment-laden flow is the main reason that causes the erosion in the turbines installed in the Himalayan Rivers. The rivers in these regions flow through newly formed Himalayas carrying sediment particles[1]. The sediment concentration in these rivers can be as high as 25,000 ppm in the wet season and completely removing such high sediment concentration is next to impossible. Quartz with hardness 7 on Moh's scale being the main ingredient comprising 70 to 95%, causes severe erosion in turbine components[2–4]. Investigation of erosion in a hydro turbine is a complex phenomenon that depends on sediment hardness, velocity, flow behaviour, turbine material properties, the shape of sediments, impingement angle, and others. The erosion leads to a change in flow behaviour which decreases the overall efficiency of the turbine. Such a change in flow behaviour may also increase the erosion rate. That makes the investigation of flow behaviour an important factor while investigating the erosion in the turbine[5–8]. I. Finnie [9] and Fuyan et.al [10] have presented the experimental results for the dependency of weight loss or erosion on velocity and impingement angle, which are shown in Figure 1 and Figure 2.

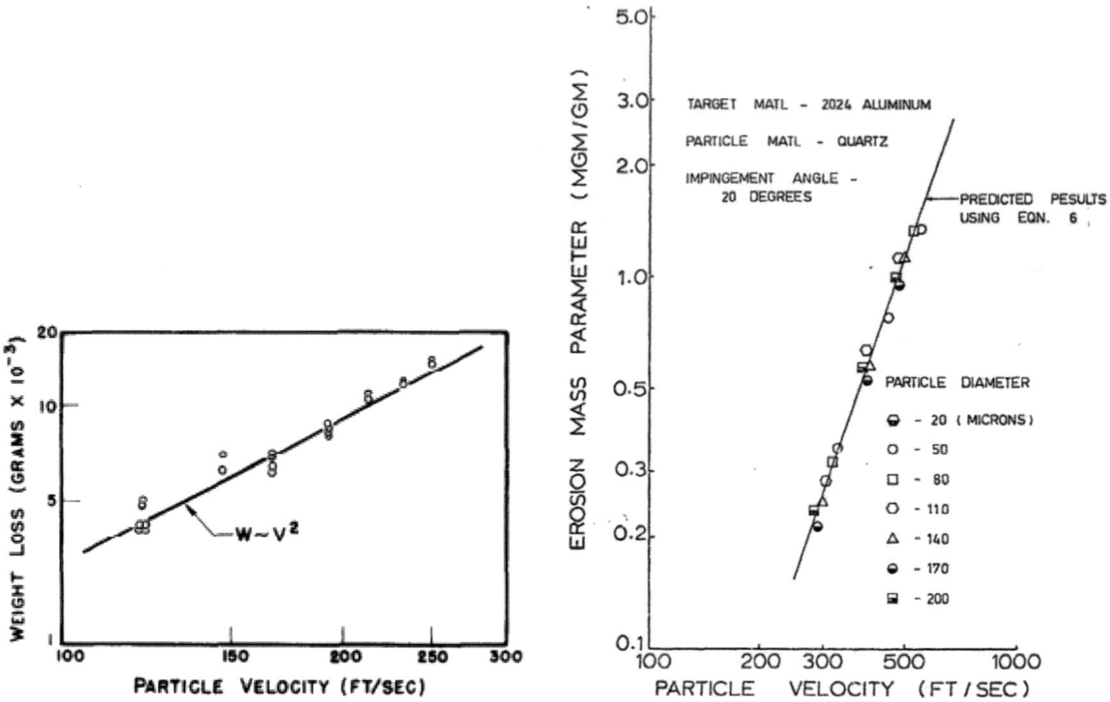

Figure 1. Dependency of weight loss on velocity [9,10]

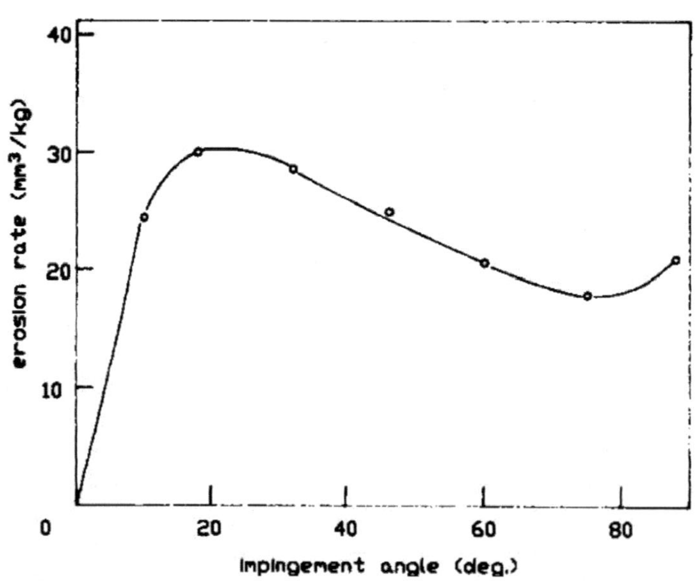

Figure 2. Dependency of erosion rate on impingement angle [9]

The operation of the Francis turbine is a complex phenomenon due to the presence of stationary as well as rotating domains. The clearance region or gap present in the Francis turbine between GV and runner blade separates the stationary GV domain and rotating RV domain as shown in Figure 3. There can be three types of side-wall clearance between guide vanes and runner i.e., guide vanes being above the level of the runner (clearance type 1), runner being above the level of the guide vanes (clearance type 2), and both the runner and the guide vanes being in the same level. In Francis turbine, operating in sediment-laden flow, erosion concentrations were observed in the region of the GV and the runner blade[11, 12]. More importantly, sever erosion was observed in side-wall clearance region of Francis runner as shown in Figure 4. The presence of height difference or clearance gap is expected to have an impact on flow behavior causing erosion in the region. This paper focuses on the comparison of flow behavior and the corresponding erosion potential for the first two clearance types (clearance types 1 and 2).

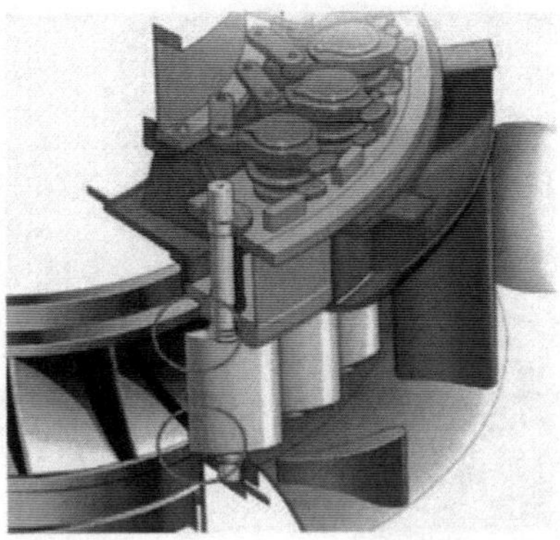

Figure 3. Sectional 3D view of Francis turbine to show side-wall clearance region [13]

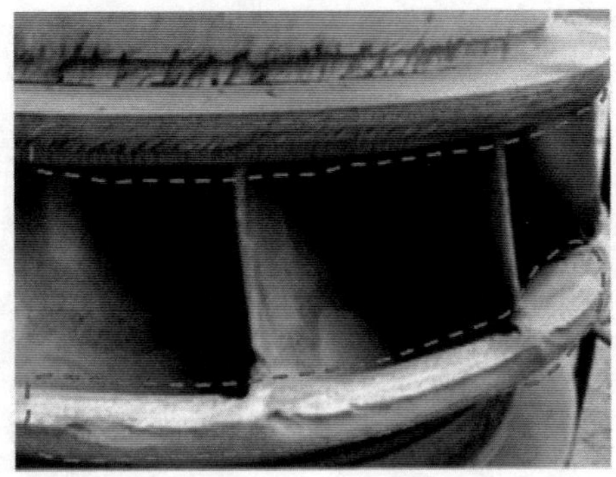

Figure 4. Erosion in side-wall clearance of Francis turbine runner [13]

2. Materials and Methods

This study comprises numerical analysis and its experimental validation. The experimental works were done using Rotating Disc Apparatus (RDA) and numerical works were done using OpenFOAM. For both experimental and numerical setup, the simplified geometries that replicate the side-wall clearance between GV and RV were created. The replication of geometries and procedure followed are explained in detail in the following sections.

2.1. *Experimental Setup*

The flow behavior and erosion potential in the side-wall clearance region for clearance types 1 and 2 only being the scope of this study, an experimental sample that replicates the region was designed as shown in the Figure 5.

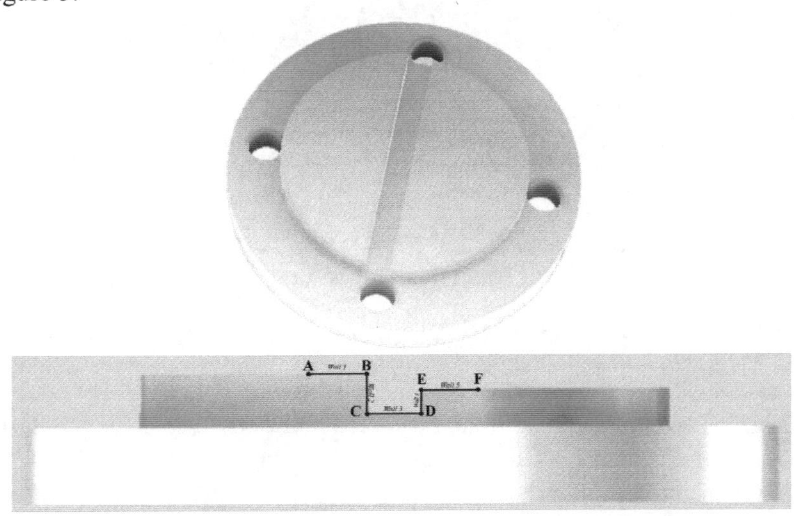

Figure 5. Experimental Sample

The BCDE section in Figure 5 resembles the side-wall clearance between GV and runner for both clearance types. The experimental sample shown in Figure 5 can be implemented for both clearance types just by changing flow direction. The clearance types with corresponding flow directions are tabulated in Table 1. The flow direction is in accordance with the experimental sample shown in Figure 5.

Table 1. Side-wall clearance types and corresponding flow directions

Types	Relative position of guide vane and runner	Flow direction
Type 1	Guide vane above the runner	Left to right
Type 2	Guide vane below the runner	Right to left

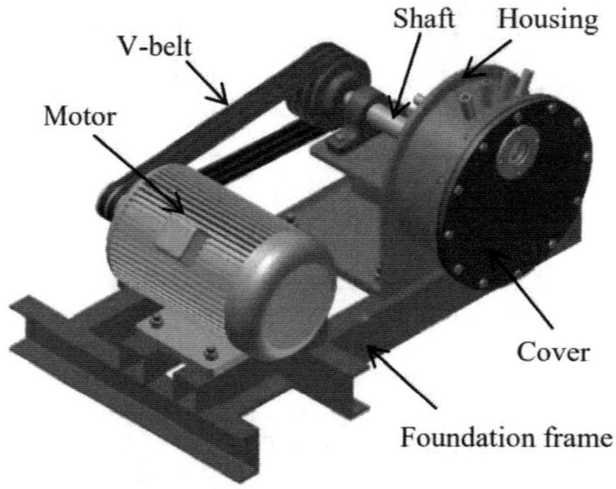

Figure 6. Rotating disc apparatus (RDA) [8]

Figure 7. Rotating disc

All the experimental works were carried out using Rotating disc apparatus (RDA). The major components of RDA are shown in Figure 6 and Figure 7. The rotating disc shown in Figure 7 is the component where the sample(s) are mounted. The hubs encircled by the red circles in Figure 7 are the places to mount the samples. The hub encircled by the white circle in Figure 7 is the place where the shaft is mounted. First of all, the experimental samples were inserted in the allocated hubs of the rotating disc. The rotating disc was then fitted inside the RDA housing and mounted over the shaft at the center such that the axis of rotation will be the center of the disc. After tightening the rotating disc, the RDA housing is closed with a cover. The calculated amount of sediment is then poured inside the RDA housing from the opening at the top of the housing. Then the hub is filled with water before closing the opening at the top. The disc is then rotated at a desired speed by an AC motor with a belt drive mechanism. The rotational speed of the motor is controlled by a variable frequency drive (VFD) machine. The VFD machine changes the frequency of the motor to change its speed. The actual rotational speed at the shaft is measured by Tachometer. The working of RDA is explained briefly in the paper [8], and the same approach was implemented for this study as well. The rotational direction of the rotating disc is responsible to achieve a particular flow direction. That is why special care should be given to the rotational direction of the disc.

The experimental sample of aluminum was used. This experimental study was done to investigate the erosion pattern in wall 2, wall 3, and wall 4 as shown in Figure 5. For that, the region BCDE was considered along wall 2, wall 3, and wall 4. The erosion in wall 2 is also affected by the erosion near the right end of wall 1. A similar condition is applied for the erosion around point E in wall 4. That is why the regions AB and EF were also considered for this study. The regions AB and EF were named wall 1 and wall 5 respectively.

2.2. Numerical Setup

The work has started with the preparation of the fluid domain. To avoid numerical complexity and the need for high computational power, the simplified fluid domain as shown in Figure 8 was adopted to have a clear look at the region of interest. The geometry was created using *onShape* CAD software. The created CAD model was then exported as *.stl* file and then the mesh was created using *snappyHexMesh*. The mesh had 768432 cells and 2380240 faces. All the cells in the mesh were hexahedral. The mesh non-orthogonality was 0.

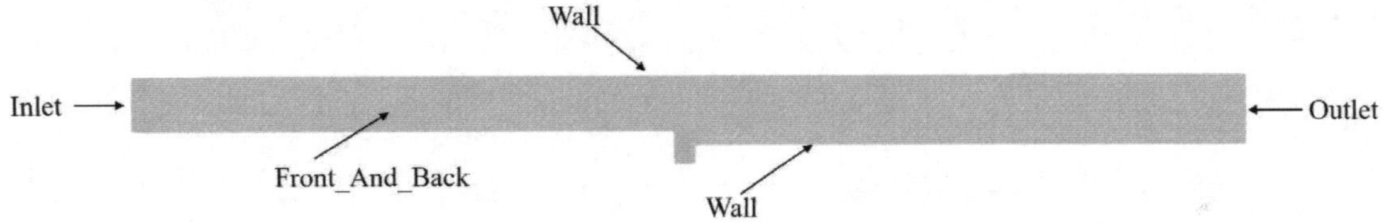

Figure 8. Boundaries in the fluid domain and their patch name

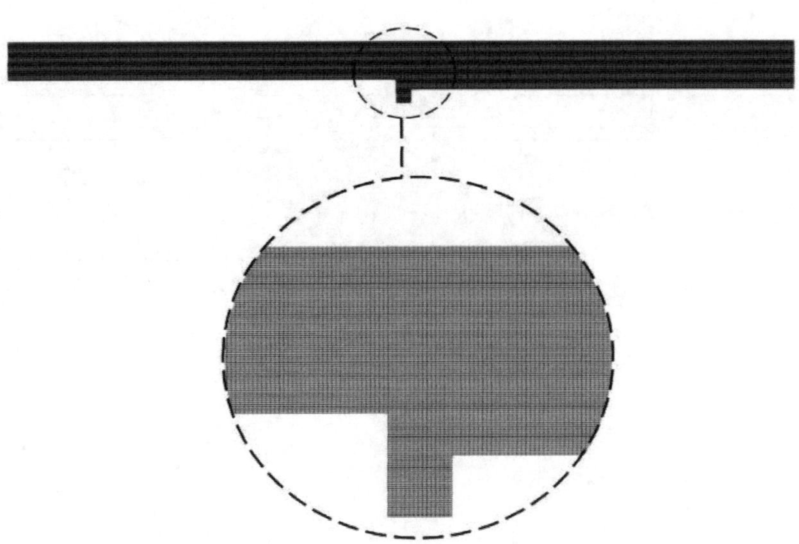

Figure 9. Meshing

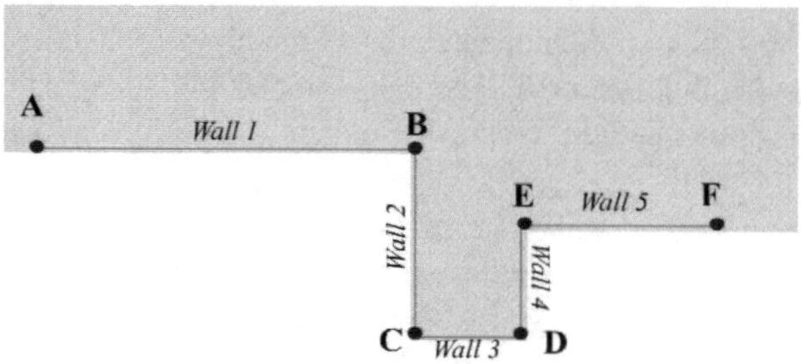

Figure 10. Region of interest

The fluid domain was taken in such a way that the length before and after the slot is 10 times the inlet width to achieve a fully developed flow before the slot. The whole computational domain is not the region of interest. As in the experimental sample, a region of interest was introduced in the computational domain as shown in Figure 10. The corresponding region of interest in the experimental sample is shown in Figure 5.

PIMPLE pressure correction algorithm which is the combination of *SIMPLE* and *PISO* pressure correction algorithm is used for the simulation of this test case. The *PIMPLE* solver is a transient solver for incompressible flow. The working flow chart of *PIMPLE* pressure correction algorithm is shown in the *Figure 11*. The discretization of the convective term in the momentum equation was done by the second-order upwind differencing scheme *(linearUpwind),* k-Epsilon transport equation by the upwind scheme *(upwind),* gradient and diffusion terms by the central differencing scheme *(liner)* and the temporal derivatives by 1st order implicit scheme.

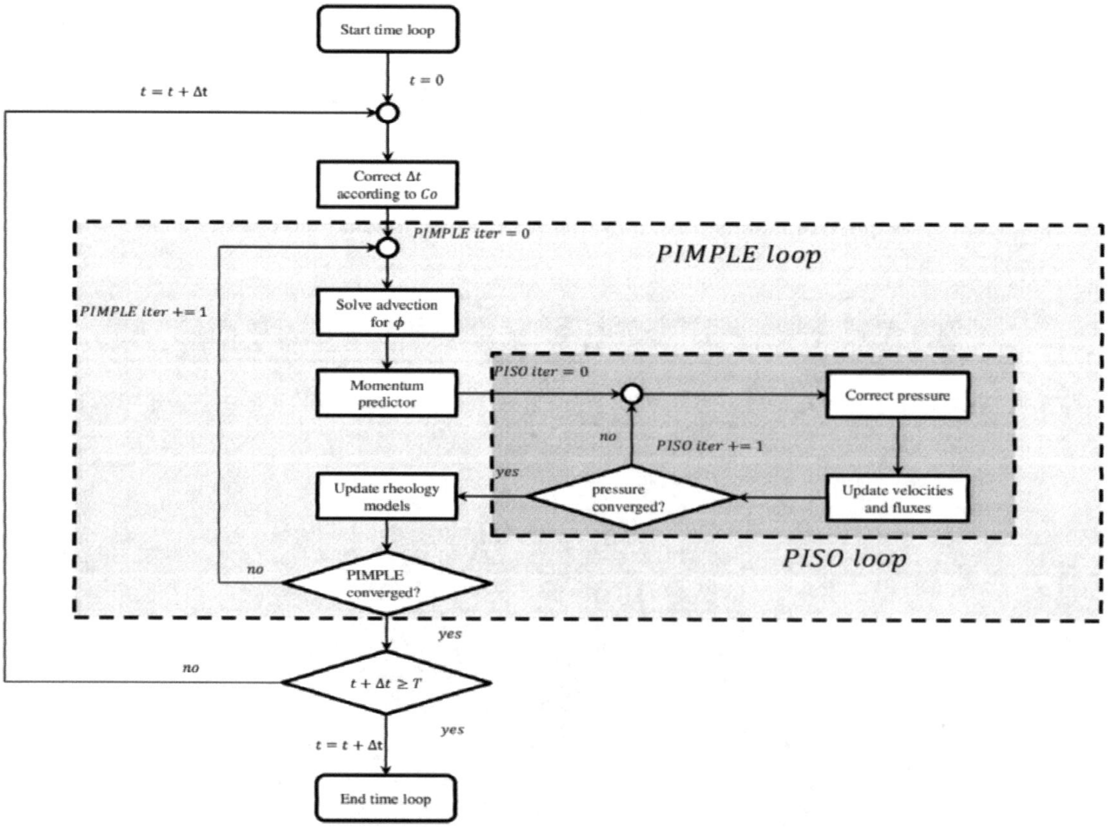

Figure 11. PIMPLE flow chart [14]

To achieve side-wall clearance type 1, the calculated inlet velocity shown in Table 2 was imposed from the left side of the geometry which corresponds to Reynold's number 3500. The same velocity was imposed from the right side of the geometry to achieve side-wall clearance type 2. The corresponding inlet velocity for a given Reynolds number was calculated using Equation (1).

$$U = \frac{R_e \times n_u}{L} \tag{1}$$

Here, U is inlet velocity, L is characteristic length, $\boldsymbol{R_e}$ is Reynolds number and $\boldsymbol{n_u}$ is kinematic viscosity. Water was taken as the working fluid during this simulation. The initial guess for k, ε and, ω was taken from Equations (2), (3) and, (4).

$$k = 1.5(U \times I)^2 \tag{2}$$

$$\varepsilon = 0.09^{0.75} \frac{k^{1.5}}{0.07 \times L} \tag{3}$$

$$\omega = \frac{\varepsilon}{k} \tag{4}$$

Here, I is turbulence intensity which is taken as 5% in this simulation, k is turbulent kinetic energy, $\boldsymbol{\varepsilon}$ is dissipation rate and ω is specific dissipation rate.

Table 2. Boundary conditions

Patches	$\boldsymbol{U}$ (ms^{-1})	$\boldsymbol{p}$ (m^{2}s)	$\boldsymbol{k}$(m^{2}s^{-2})	$\boldsymbol{\varepsilon}$(m^{2}s^{-3})	$\boldsymbol{\omega}$(s^{-1})
Inlet	fixedValue, uniform (0.4 0 0)	zeroGradient	fixedValue, uniform 0.0006	fixedValue, uniform 0.00345	fixedValue, uniform 5.74
Outlet	zeroGradient	fixedValue, uniform 0	zeroGradient	zeroGradient	zeroGradient
Wall	noSlip	zeroGradient	kqRWallFunction, uniform 0.0007	epsilonWallFunction, uniform 0.00345	omegaWallFunction, uniform 5.74
Front_And_Back	empty	empty	empty	empty	empty

The run time for the simulation was set for 0.3 seconds with a time-step of 0.001 seconds. The courant number remained below 1 for most time during the simulation. The k-Omega turbulence modeling was adopted for this study. Some particles were injected from the inlet in the flow to look after the particle motion in ParaView. The graphs for impingement angle and velocity was plotted along the region of interest to predict the erosion locally in the region.

3. Results and Discussions

3.1. Experimental Results

The reference points ABCDEF and the corresponding walls numbering to make a comparison in erosion pattern between clearance types 1 and 2 are taken from Figure 5. The side-wall clearance gap region BCDE comprises wall 2, wall 3 and, wall 4. For clearance type 1, some faint erosion can be seen at two ends of wall 2 i.e., around points B and C whereas maximum erosion was observed around these points for clearance type 2. Not only the endpoint region but the complete wall 2 is the most eroded wall in case of clearance type 2 but it is the least eroded wall in case of clearance type 1. The region around points D and E are highly eroded for clearance type 1 but not for clearance type 2. A similar erosion pattern was seen throughout the depth of the slot for clearance type 2 but an unusual erosion was observed at the far end of wall 3 for clearance type 1. Wall 3 is the most erosion-prone wall and wall 2 is the least erosion-prone wall for clearance type 1, and wall 2 is the most erosion-prone wall and wall 4 is the least erosion-prone wall for clearance type 2.

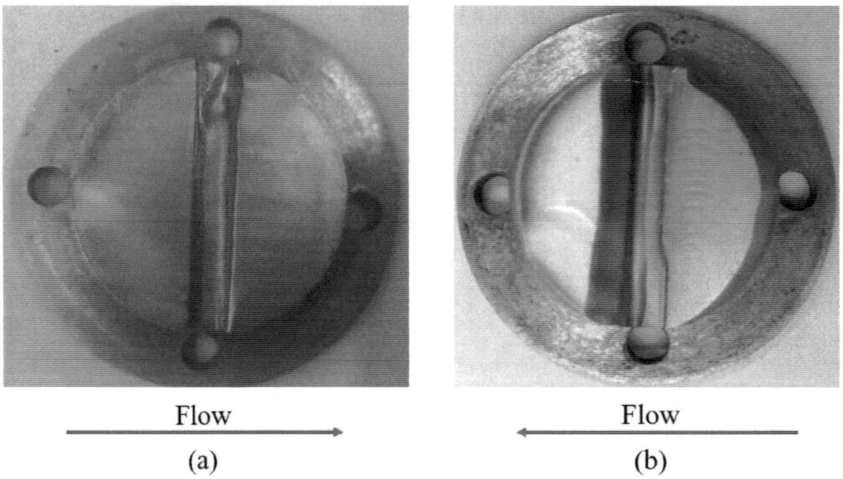

Figure 12: Experimental result for side-wall clearance (a) type 1, (b) type 2

3.2. Numerical Results

The velocity contours in Figure 13 show that the velocity at the middle portion of the slot is less than other parts of the contour for both clearance types. When specifically looking at the slot region BCDE, the velocity near the walls 2, 3, and 4 is higher than that in the middle portion of the slot for both clearance types. The near wall velocity in the slot region in clearance type 2 is much higher than in clearance type 1. This high near wall velocity is the reason for the high erosion in clearance type 2 as suggested by the graph in Figure 1.

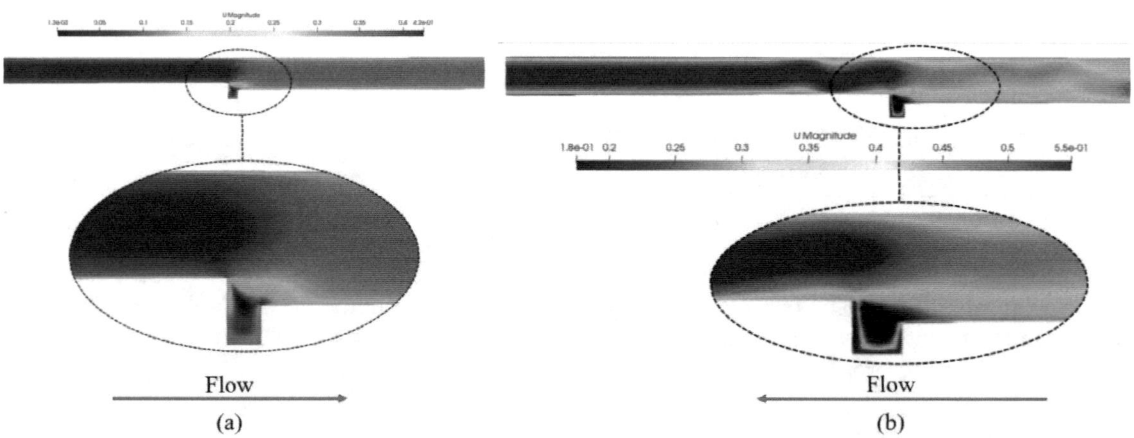

Figure 13. Velocity contours for clearance (a) type 1, (b) type 2

The impingement angle and velocity were plotted along the pre-defined region of interest for both clearance type for better comparison of the erosion in each wall. The Figure 2 shows the erosion rate is maximum for impingement angle 10^0 to 45^0 and Figure 1 shows the erosion rate increases with velocity. These two relations were considered to analyze different erosion pattern at different location around the region. For that, each wall was made comprising 1000 points. The velocity and impingement angle were calculated at each point to analyze the erosion pattern.

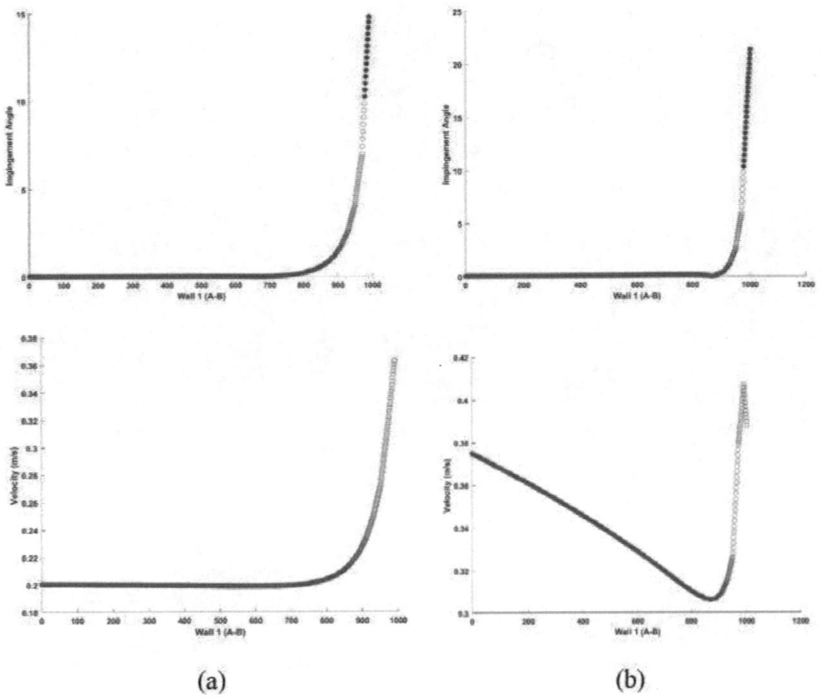

Figure 14. Impingement angle and velocity in wall 1 for clearance (a) type 1 (b) type 2

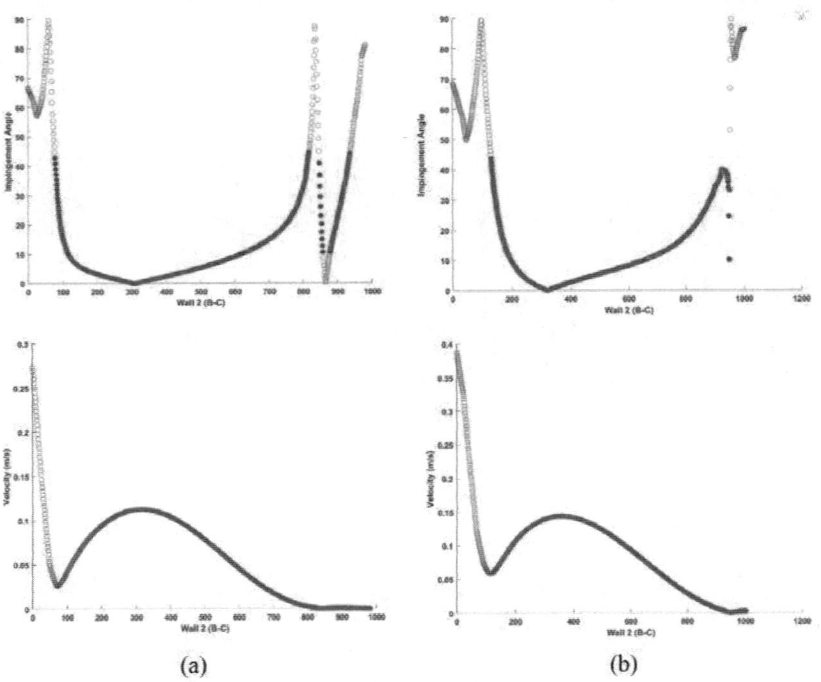

Figure 15. Impingement angle and velocity in wall 2 for clearance (a) type 1 (b) type 2

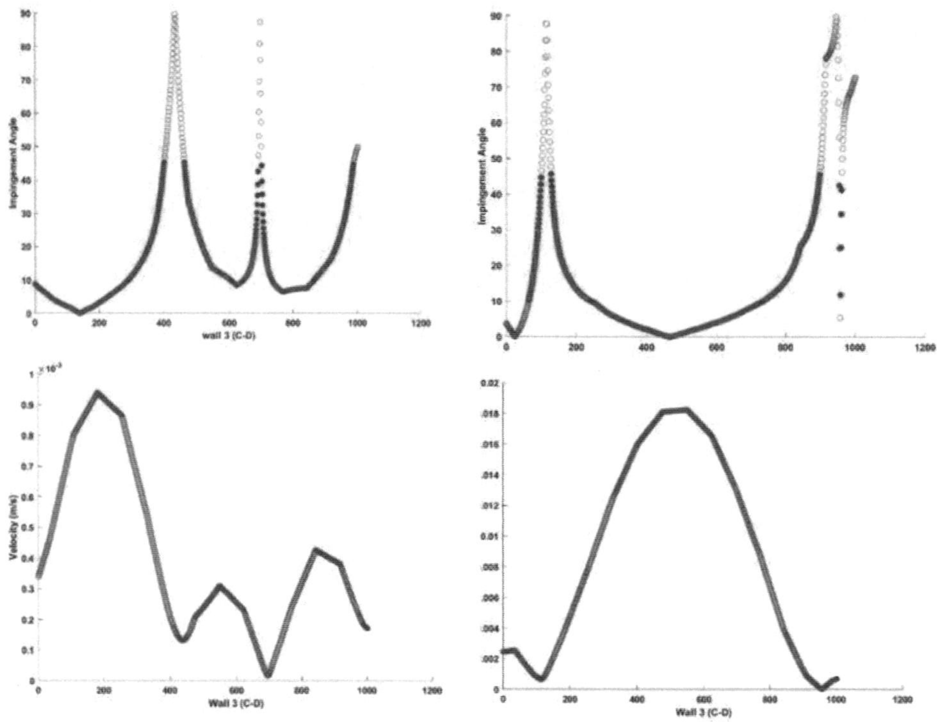

(a) (b)

Figure 16. Impingement angle and velocity in wall 3 for clearance (a) type 1 (b) type 2

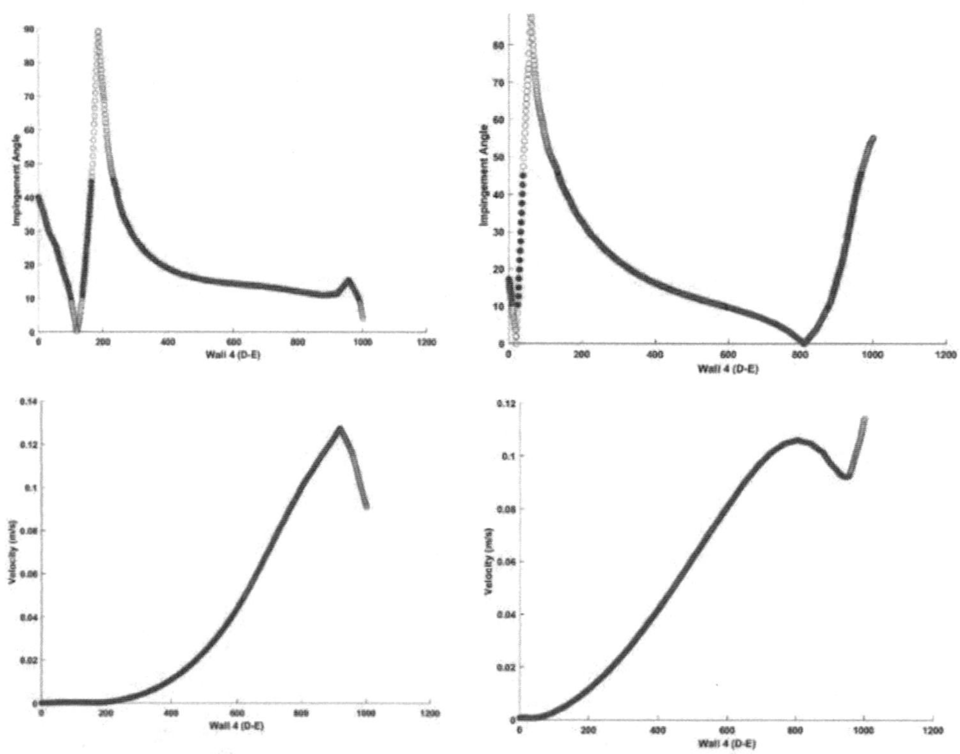

(a) (b)

Figure 17. Impingement angle and velocity in wall 4 for clearance (a) type 1 (b) type 2

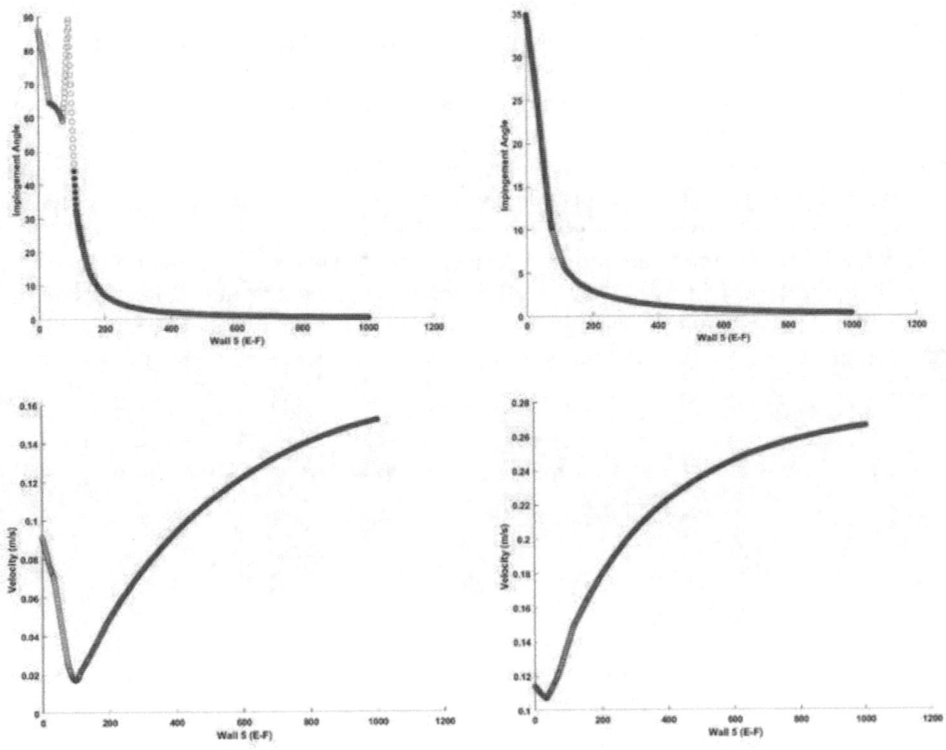

(a) (b)

Figure 18. Impingement angle and velocity in wall 5 for clearance (a) type 1 (b) type 2

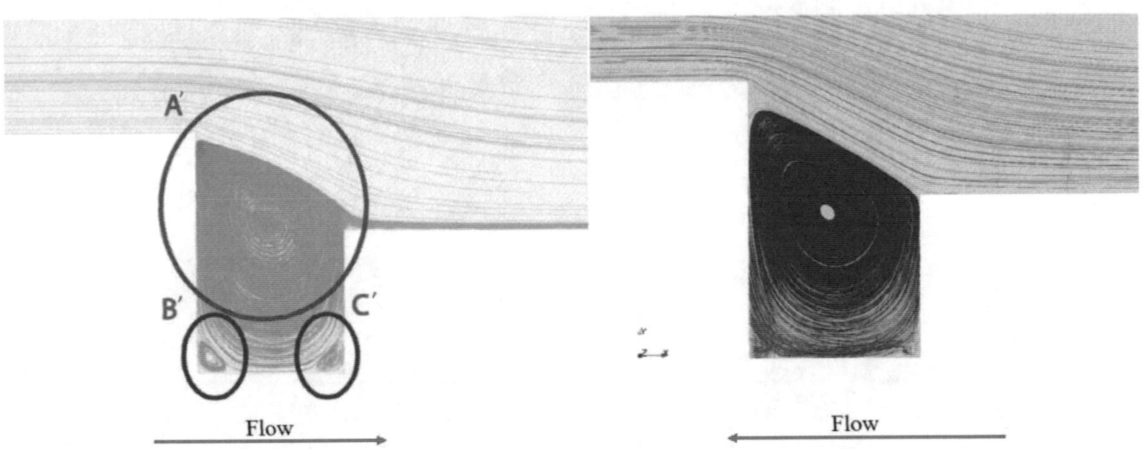

(a) (b)

Figure 19. Streamlines for clearance (a) type 1, (b) type 2

The red zone in the above graphs denote the area where the impingement angle is between 10^0 and 45^0, where the erosion rate is higher as discussed earlier. For wall 1, the graphs of impingement angle looks quite similar for both clearance types, i.e., only some region around point B falls in the red zone of impingement angle. The actual difference can be seen in the velocity plot as the velocity of whole region AB is greater than 0.3 for clearance type 2 whereas that maximum velocity of around 0.3 can be seen for clearance type 1. The region around point B for both clearance types falls both in high velocity zone and red zone of impingement angle. That's why the region around point B is more prone to erosion for both clearance types as far as the wall 1 is concerned. The velocity in the red zone region is higher for the clearance type 2, that's why comparatively high erosion can be seen for clearance type 2 than that in clearance type 1. For wall 2, the end point regions, i.e., regions around point B and C falls in the red zone of impingement angle for both clearance types. The corresponding velocity of red zone regions is much higher for clearance type 2. For wall 3, the velocity is comparatively lesser for both clearance types. Almost whole region except region around point C falls under the red zone of impingement angle for clearance type 2 whereas the far ends of wall 3 only falls in the red zone region for clearance type 2. For wall 4, the range of velocity is quite similar for both clearance types. Almost whole domain falls under the red zone of impingement angle for clearance type 1. That's way comparatively high erosion was seen for clearance type 1 in wall 1. For wall 5, the region around point E falls in the red zone of impingement angle. The velocity in clearance type 2 was found to be higher than that in clearance type 1. For both clearance types, erosion around point E is more dominant as far as wall 5 is concerned.

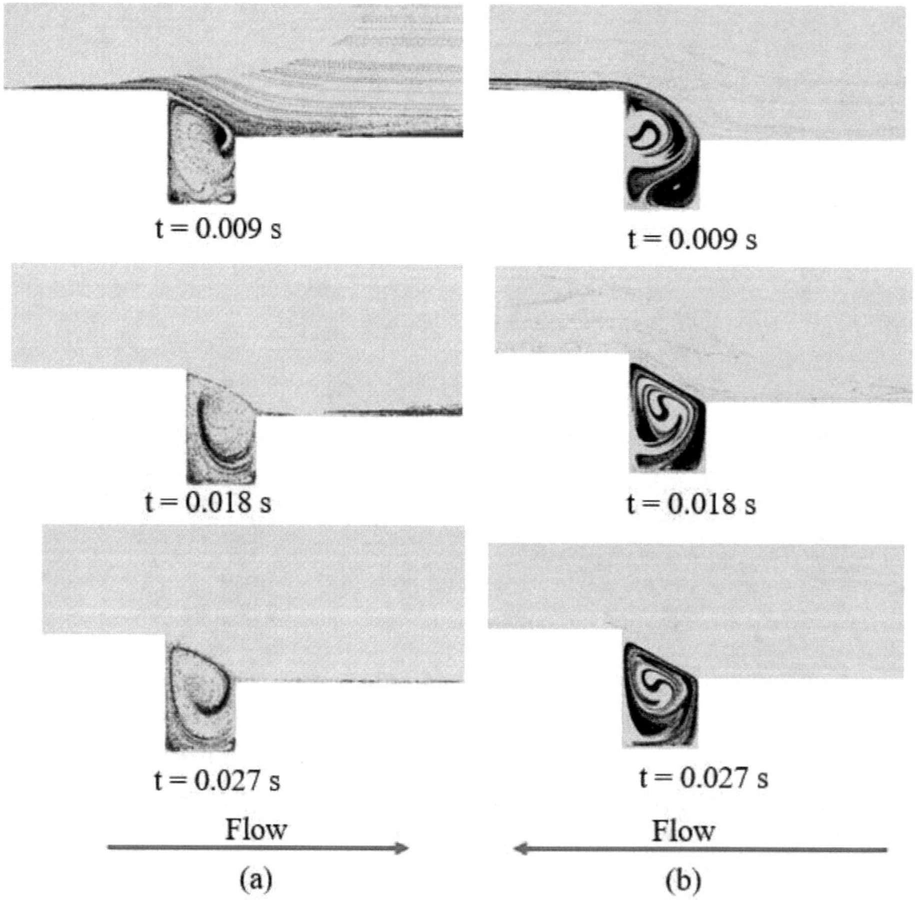

Figure 20. Particle flow for clearance (a) type 1, (b) type 2

The streamlines in Figure 19 show the formation of vortices in the slot region but for clearance type 1, along with large vortex A', the small vortices B' and C' are also formed at the corners. The formation of vortices is the main reason for the low velocity at the slot region. Some particles were injected into the flow to have closer look at the behavior of particles in the vortex and non-vortex regions as shown in Figure 20. The particle motion shows that these particles hit the opposing edge i.e., point E in case of clearance type 1 and point B in case of clearance type 2. The erosion was also observed around these regions during the experiment. The particles also formed vortices in the slot region. The formation of the vortices made the particles revolve in the slot hitting the edges and faces of the slot. Even after stopping the injection of particles, the particles trapped in the slot did not escape. This indicates that if sediment particles once get trapped in the slot, then they keep on hitting the edges and faces of the slot causing erosion in the region. The trapped particle can only get released from the slot by causing erosion in the region. The velocity near the walls of the slot is higher for clearance type 1 than that for clearance type 2, this makes the trapped sediment particles hit walls with high velocity for clearance type 2. That is why high erosion was observed for clearance type 2 in experimental results. The formation of vortices B' and C' are the reasons behind the erosion in the corresponding corners in clearance type 1. The erosion in corner C for clearance type 2 is not well predicted by the numerical analysis.

4. Conclusion

For clearance type 1, maximum erosion was observed in wall 3 and the corner between walls 3 and 4 i.e., around point D. The region BC which corresponds to wall 2 was the less erosion-prone region. The unusual erosion pattern observed at one end of wall 3 was not predicted by the numerical analysis. Only 2D numerical analysis was done, which might be the reason behind not predicting that unusual erosion pattern. It is believed that the application of 3D numerical analysis can predict that erosion if that erosion was not due to any experimental error. For clearance type 2, the maximum erosion was observed in wall 2 and the corner between walls 2 and 3. The analysis of impingement angle and corresponding velocity was found to be critical to predict erosion at local level. No unusual erosion pattern was observed in the case of clearance type 2. For both clearance types, corners and edges are more affected except that unusual erosion was observed in the face of wall 3 for clearance type 1.

The velocity in the middle portion slot region was comparatively lower than in other parts computational domain. This is due to the formation of the vortex in that region. The vortex formation in sediment-laden flow made sediment revolve in the slot hitting the walls. It was found that even after stopping the injection of sediment, the sediment which was trapped in the slot could not escape. This phenomenon has concluded that little sediment in the slot is enough to erode the slot region as that sediment keeps on revolving in the slot hitting walls that cannot escape without eroding the wall.

5. Acknowledgements

This work is supported by the **FranSed Project** which is currently running with the collaboration of the Norwegian University of Science and Technology and, Kathmandu University. The computational support for the numerical analysis is provided by the **Kathmandu University Supercomputer Centre**.

6. References

[1] Thapa B, Sand erosion hydropower, (2004).

[2] Thapa B, Shrestha R, Dhakal P & Thapa B S, Problems of Nepalese hydropower projects due to suspended sediments, *New Pub Michigan State Univ. Press*, 8 3 (2007) 251.

[3] Padhy M K & Saini R P, A review on silt erosion in hydro turbines, *Renew. Sustain. Energy Rev.*, 12 7 (2008) 1974.

[4] (PDF) River sedimentation challenges for hydropower development in Nepal: A perspective review, https://www.researchgate.net/publication/334466363_River_sedimentation_challenges_for_hydropower_development_in_Nepal_A_perspective_review, 18 April 2022.

[5] Eltvik M, Sediment Erosion in Francis Turbines, (2009).

[6] Padhy M K & Saini R P, Effect of size and concentration of silt particles on erosion of Pelton turbine buckets, *Energy*, 34 10 (2009) 1477.

[7] Thapa B S, Thapa B & Dahlhaug O G, Empirical modelling of sediment erosion in Francis turbines, *Energy*, 41 1 (2012) 386.

[8] Rajkarnikar B, Neopane H P & Thapa B S, Comparative study of sediment erosion on alternative designs of Francis runner blade, *Int. J. Fluid Mach. Syst.*, 8 3 (2015) 183.

[9] Finne I, Erosion of surfaces, *Wear*, 3 (1960) 87.

[10] Fuyan L & Hesheng S, The effect of impingement angle on slurry erosion, *Wear*, 141 2 (1991) 279.

[11] Noon A A, Kim M-H, Anagnostopoulos J & Cimbala J M, Sediment and Cavitation Erosion in Francis Turbines—Review of Latest Experimental and Numerical Techniques, *Energies 2021, Vol. 14, Page 1516*, 14 6 (2021) 1516.

[12] Thapa B S, Dahlhaug O G & Thapa B, Effects of sediment erosion in guide vanes of Francis turbine, *Wear*, 390–391 (2017) 104.

[13] Chitrakar S, Neopane H P & Dahlhaug O G, Secondary Flow and Sediment Erosion in Francis Turbines DISSERTATION ON SECONDARY FLOW AND SEDIMENT EROSION IN FRANCIS TURBINES In Partial Fulfillment of the Requirements for the Doctorate Degree in Mechanical Engineering Sailesh Chitrakar June , 2018, June (2018).

[14] Municchi F, Nagrani P P & Christov I C, A two-fluid model for numerical simulation of shear-dominated suspension flows, *Int. J. Multiph. Flow*, 120 (2019) 1.

IOP Conf. Series: Materials Science and Engineering 1279 (2023) 012010 doi:10.1088/1757-899X/1279/1/012010

Simulation of passive thermal retrofitting techniques in an earthquake-resistant building: A case study of a building in Thakle Namuna Basti, Melamchi, Nepal

Malesh Shah[1*], **Parth Singh Devkota**[1], **Gaurav Timilsina**[1], **Dikshant Shrestha**[1], **Sirapa Shrestha**[1], **Bivek Baral**[1]

[1] Department of Mechanical Engineering, Kathmandu University, Dhulikhel, Nepal
*Corresponding Author: malesh.shah@ku.edu.np

Abstract. Thermal comfort is a relatively emerging topic in the Nepalese context. The role of thermal comfort in building performance is foreshadowed by peoples' emphasis on only the structural aspect and active heating/cooling systems. This paper analyzes the effectiveness of passive thermal retrofitting techniques in an earthquake-resistant building in a small community called Thakle Namuna Basti of Melamchi, Nepal. Only three retrofitting techniques are used in this simulation; the addition of gypsum plaster on the wall, the use of foam on the floor, and the application of double-glazed windows. The study was conducted through a two-phased survey, and simulation in TRNSYS. The heating and cooling loads before and after retrofitting are compared. It showed that applying passive thermal retrofitting techniques reduced the overall annual heating load by 35% in the winter but increased the overall cooling load by 4% in the summer.

1. Introduction

Aside from significant human casualties in the massive earthquake of 2015 in Nepal [1], many inhabited buildings which were not structurally sound were destroyed beyond recovery [2]. The building codes formulated by the Government of Nepal (GoN) did not focus sufficiently on structural aspects and were then revised [3]. In Nepal, the Department of Urban Development and Building Construction (DUDBC) regulates and implements building construction regulations through 'NBC: 105: 2020 Seismic Building Design' [4]. The building code focuses mainly on structural aspects such as load bearing, building construction materials, and architectural guidelines aimed toward the ability to withstand seismic waves and earthquakes [4]. But the building code never focused on the building's thermal aspect, its thermal performance, and the result it would have on the residents' overall comfort. Due to the presence of seasonal and climatic variations across the country, thermal comfort is a significant parameter for building construction [5]. Despite the recognition of the importance of renewable energy, energy efficiency, and energy-saving in buildings by the GoN, limited focus has been given to human comfort in terms of thermal features. Although the varying perception of thermal comfort among people adds to the complexity of the building design, its inclusion is a mandatory step in making buildings more habitable [6]. This paper conducts research on a building constructed following the DUDBC guidelines, located at Thakle Namuna Basti, Melamchi, Nepal, and incorporates passive thermal retrofitting techniques using TRNSYS. The remaining section of this paper consists of site selection, simulation methodology, and comparisons between the results prior to installing passive thermal retrofitting techniques, and the results after integrating them.

IOP Conf. Series: Materials Science and Engineering 1279 (2023) 012010 doi:10.1088/1757-899X/1279/1/012010

2. Site Selection

Established community villages were considered potential sites for this study due to the convenience of gathering concentrated sources of information. Further, all the buildings have similar physical structures, so any building can be randomly selected for analysis. Selection among those sites was done through a survey in two phases which are described below.

2.1. Survey: First Phase

The first phase of the survey was carried out in Thakle Namuna Basti at Thakle, Dhurmus Suntali Namuna Basti at Giranchaur, and Laprak Namuna Basti at Laprak. The following parameters were chosen as selection criteria among the three sites:

2.1.1. Structural Soundness. The building must be structurally sound to ensure it can withstand additional physical modification after deciding on appropriate retrofitting techniques for thermal comfort. Further, for this study, the building constructed following DUDBC guidelines is preferred.

2.1.2. Construction Completeness. In case the construction of the building is not complete, there is a possibility of the addition of structures with materials not considered in the study which can cause deviation in thermal comfort.

2.1.3. Inhabitation Status. Conducting a study on an uninhabited building would be useless and assuming thermal comfort in the building from the survey of nearby residents makes the findings inaccurate.

2.1.4. Location Appropriateness. For this study, a site that is easily accessible with proper transportation infrastructure is chosen.

2.1.5. Test Equipment Installation Feasibility. Data loggers had to be installed in the building to record the temperature. As the residents were not well informed about the technology and significance of the study, installing them for a prolonged period required their willingness to participate.

2.1.6. Passive Retrofitting Applicability. After coming up with the modifications required to increase thermal comfort in the building, the habitants would have to agree upon installing them. Financial and awareness-related constraints can hinder the application of the findings of the study.

Table 1 shows the status of the three potential locations based on the selected criteria. The community village of Thakle fulfilled all the criteria. Additionally, Giranchaur and Laprak did not follow DUDBC guidelines for building construction. Further, the community village of Laprak did not satisfy other criteria as well including construction completeness, inhabitation status, location appropriateness, and test equipment installation feasibility. So, Thakle Namuna Basti was chosen as the site for the study.

Table 1. Criteria for site selection.

Criteria	Thakle	Giranchaur	Laprak
Structural Soundness	Yes	No	No
Construction Completeness	Yes	Yes	No
Inhabitation Status	Yes	Yes	No
Location Appropriateness	Yes	Yes	Yes
Test Equipment Installation Feasibility	Yes	Yes	No
Passive Retrofitting Applicability	Yes	Yes	Yes

2.2. Survey: Second phase

The second phase of the survey was carried out at Thakle Namuna Basti (picture shown in Figure 1). First, information regarding environment setting, altitude, temperature, orientation, year of construction, and building design was collected and is shown in Table 2. South-faced buildings were considered for further analysis. Materials used in the buildings for different structural components were also collected. They are tabulated in Table 3.

Table 2. General building description of Thakle Namuna Basti.

Title	Description
Environment Setting	Rural
Altitude	1390 m.a.s.l
Average Annual Temperature	18.1 °C
Building Orientation	North and South Faced
Year of Construction	2020
Building Design	1.5 Storeyed, 9 Windows, 6 Doors

Table 3. Materials used in different components of the building.

Building Components	Materials
Ground Flooring	Concrete
Structural Wall	Concrete Block, Plaster (Inside)
Roofing	Zinc Sheet
Window	Single-Glazed Glass
Doors	Wood

The residents of the community were interrogated by developing a questionnaire taking reference to the American Society of Heating, Refrigerating and Air-Conditioning Engineers (ASHRAE) scale of thermal comfort. The questionnaire included questions regarding the level of comfort the residents felt in different seasons (scale of 0-5 from extreme cold to extreme hot perception), coping methods to achieve thermal comfort (fan/heater, cold/hot beverage, clothing, window blinds, door openings, and air vent), the main reason for thermal discomfort (humidity, air movement, direct sunlight and rigid clothing habit), clothing preference, effect in work efficiency due to thermal discomfort, willingness to retrofit building (changing building envelop, internal partition, shading, insulation, number of windows and windows/doors material) and agreeable range of investment. Additionally, building dimensions were also measured for further analysis.

Figure 1. Picture of Thakle Namuna Basti, Melamchi, Nepal

3. TRNSYS Simulation

TRNSYS software was used to simulate thermal performance by incorporating passive thermal retrofitting techniques. The simulation was carried out in the following steps:

3.1. Defining the house in TRNBuild

The house constructed at the selected site is one and a half storeyed. It consists of 3 bedrooms, a kitchen, and a bathroom on the first floor. The 3 bedrooms are designated as zone 1, zone 2, and zone 5, the kitchen as zone 4, and the bathroom as zone 3. The floor plan of the first floor is shown in Figure 2(a). The second floor consists of a small attic used for storage. The entire space of the second floor is designated as zone 6. The floor plan for the first floor is shown in Figure 2(b).

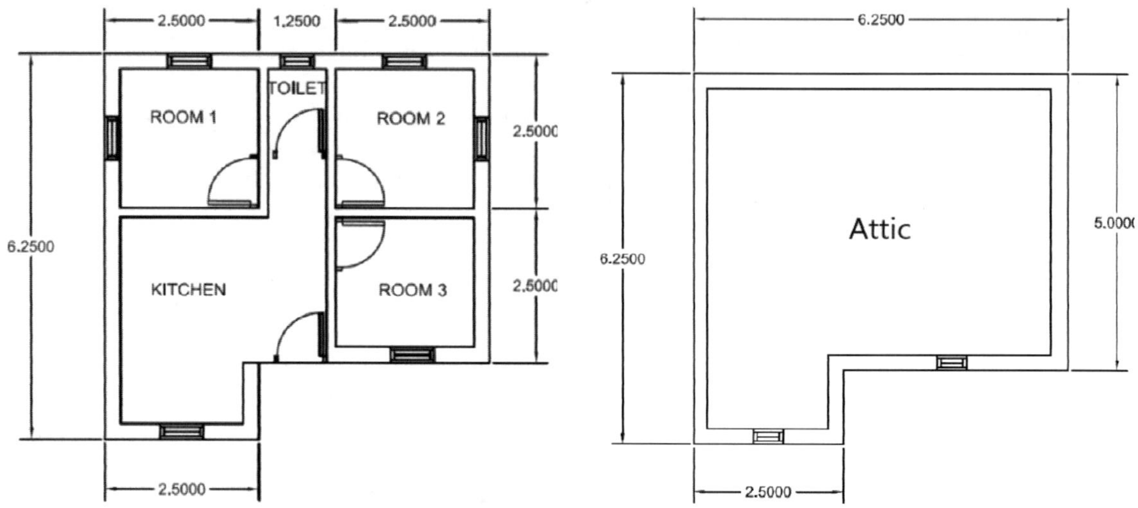

(a) **(b)**

Figure 2. Shown (a) are the dimensions for the first floor and shown (b) are the dimensions for the attic.

An initial model was built in TRNBuild after defining the zones. The floor area, volume, and surface area of each zone were fed as input as given in Table 4.

Table 4. Floor area, volume, and surface area for six zones.

	Floor area (m^2)	Volume (m^3)	Surface area (m^2)
Zone 1	6.25	21.87	41.25
Zone 2	6.25	21.87	41.25
Zone 3	1.25	4.37	17.00
Zone 4	14.37	50.31	77.39
Zone 5	6.25	21.87	41.25
Zone 6	34.37	68.75	125.38

The next step was to define materials for each structural component of the building. The ground flooring was made of concrete, the structural walls were made of interlocking concrete blocks with plaster and paint on the internal side, the roof was made with zinc sheet, the windows with single glazed glass, and the doors were made of wood. Table 5 shows the materials with their corresponding thermal conductivity, density, specific heat capacity, and overall U-value given by TRNBuild. The partition of the floors was made of plywood. The main structure of the building was made of iron bars, but it was not considered in the study as they contribute less to the overall thermal change in the house.

Table 5. Materials used to define initial building in TRNBuild with their properties.

Building Components	**Materials**	**Thermal Conductivity (W/mK)**	**Density (kg/m^3)**	**Specific Heat Capacity (J/kgK)**	**Overall U Value (W/m^{2}K)**
Ground Flooring	Concrete	1.13	2000	1000	4.30
Structural Wall	Concrete Block	1.13	2000	1000	4.30
	Plaster (Inside)	1.20	72	9200	2.53
Roofing	Zinc Sheet	50.00	7000	388	6.83
Window	Single-Glazed Glass	0.96	2500	790	5.80
Doors	Wood	0.13	600	1760	2.27

Three passive retrofitting techniques were chosen to increase thermal comfort in the building. A massless foam was added to the ground's flooring, the initial plaster was replaced by gypsum plaster in the structural walls and the single-glazed glasses were replaced by double-glazed glasses in the windows. A new retrofitted model was then built in TRNBuild. Table 6 shows the materials used for passive retrofitting and their corresponding thermal conductivity, density, specific heat capacity, and overall U-value given by TRNBuild. A glimpse of data addition in TRNBuild is given in Figure 3.

Table 6. Additional retrofitting materials with their properties.

Building Components	Material	Thermal Conductivity (W/mK)	Density (kg/m^3)	Specific Heat Capacity (kJ/kgK)
Ground Flooring	Massless Foam	0.024	35	1.47
Structural Wall	Gypsum Plaster	0.35	1200	0.84
Window	Double-Glazed Glass	0.04	-	-

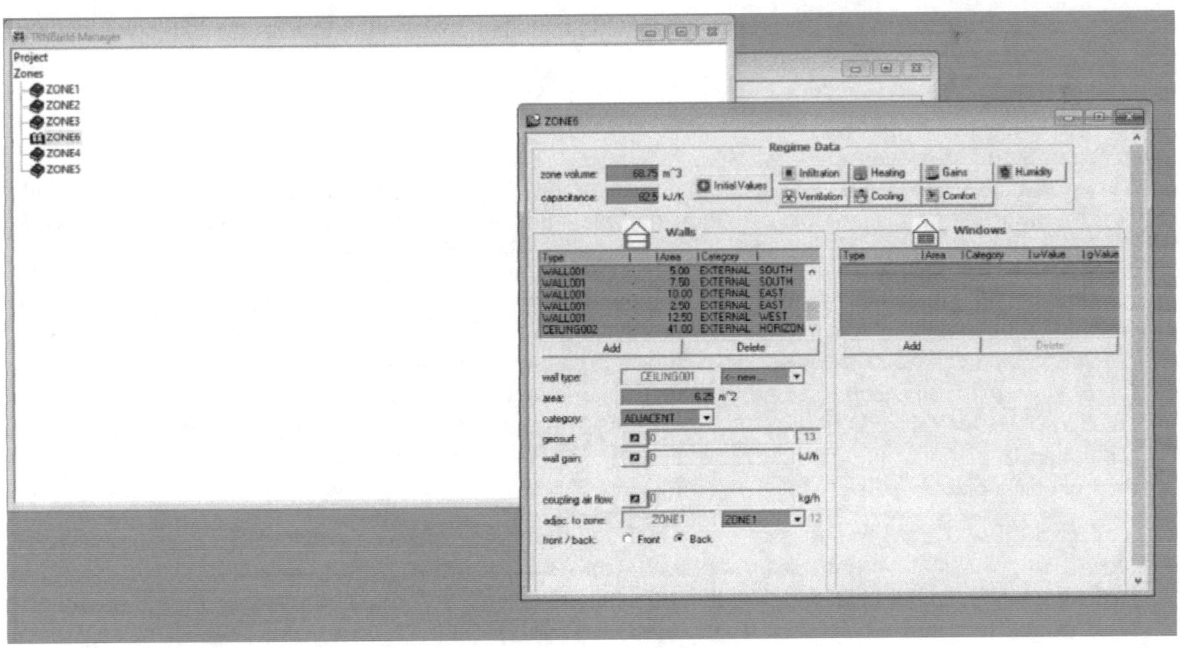

Figure 3. The zones and their relevant dimensions defined in TRNBuild.

3.2. Defining the Energy Model in TRNSYS Studio

After defining the house in TRNBuild, energy modelling was done using TRNSYS Studio. The defined house in TRNBuild was inserted in the TRNSYS Studio via the building component 'Type 56b' [7]. Type 56b in TRNSYS is a balancing model with non-geometric features with each zone having only an air node [7]. A file describing the building's characteristics is fed into and processed by the PREBID software and then utilized during simulation by the type 56b component [7].

Weather data from Meteonorm 8.0 and the residents' schedules were fed, and the required output was extracted from the building component. The model constructed in TRNSYS Studio is shown in Figure 4. Using the model, energy consumption for the house without retrofitting techniques and with retrofitting techniques was determined. The results are compared in the results section of this paper.

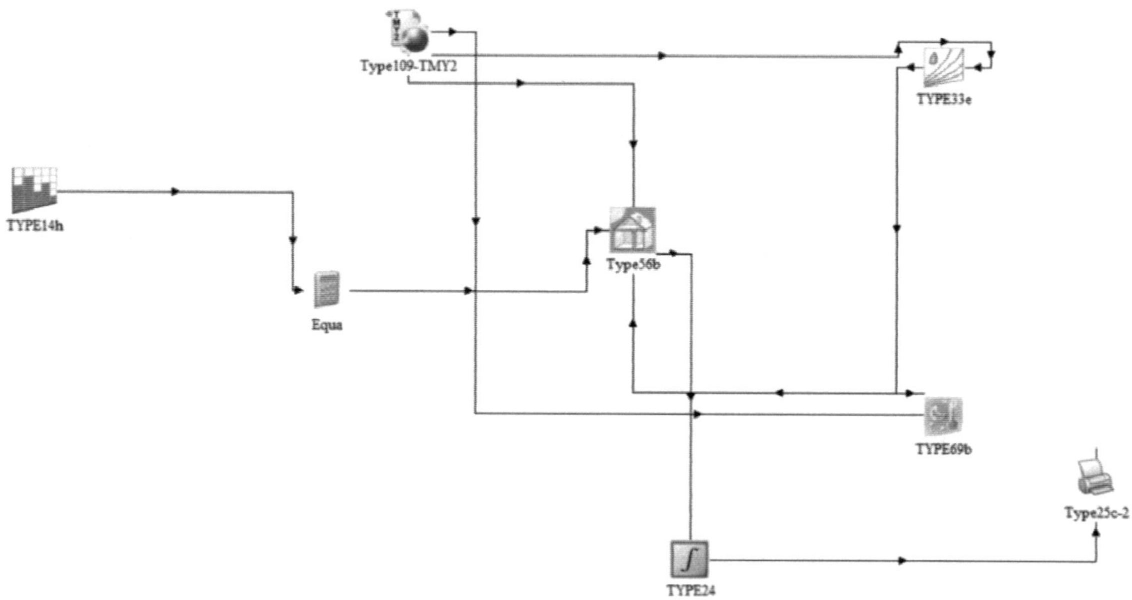

Figure 4. The energy modelling of the house using TRNSYS Studio.

4. Results and Discussion

The first phase survey showed that Thakle Namuna Basti had structurally sound buildings whose construction had been completed by the time this study was done. They were regularly inhabited by residents and were in a location with proper transportation infrastructure.

The second phase survey was done to know the thermal comfort perception of the residents and showed that they felt extremely cold during the winter season. However, the climate during the summer season was perceived as pleasant. They did not use electrical appliances such as fans and heaters to increase their thermal comfort due to their financial status. Their clothing habits resembled common patterns observed throughout the country. Thermal discomfort hampered their working schedule during the winter (e.g., household work). They showed readiness to retrofit their building thermally given that they get financial aid entirely from external sources.

TRNSYS was used to compare the initial conditions without any retrofitting techniques and the retrofitted conditions with the incorporation of massless foam for ground flooring, gypsum plaster for structural walls, and double-glazed glasses for windows. Figure 5 shows cooling loads before and after retrofitting, for a year as obtained from TRNSYS Studio for the six zones, and Table 7 shows changes in cooling loads in different zones before and after retrofitting. Cooling load decreased by approximately 0.98 times in the attic while it increased in the other zones with the maximum increase in the bathroom. Overall the cooling load in the summer increased by 1.04 times when all zones were considered.

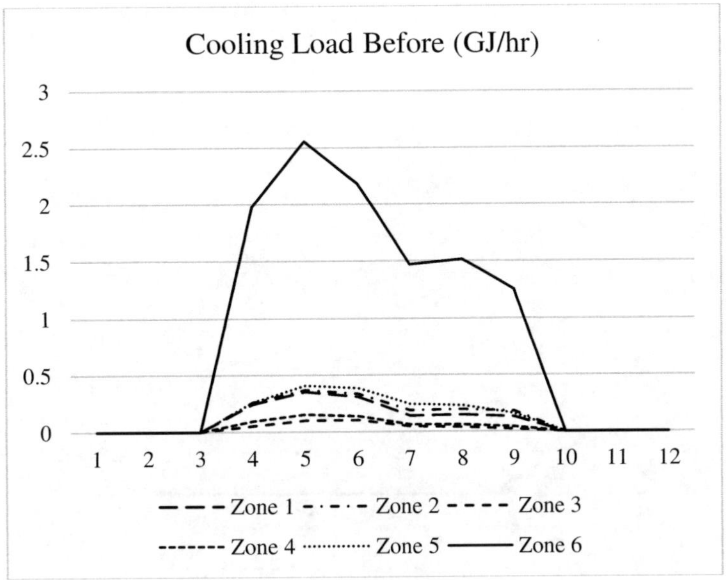

(a)

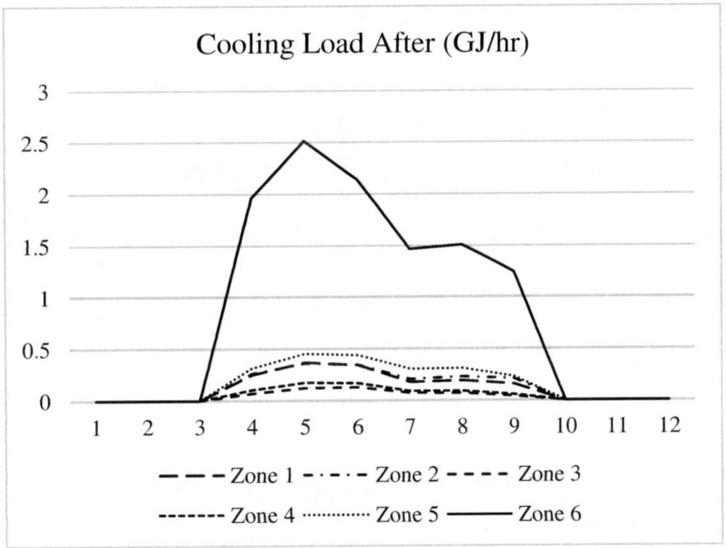

(b)

Figure 5. (a) Monthly cooling load before retrofitting (b) Monthly cooling load after retrofitting

Table 7. Change in cooling load after retrofitting.

Zone	Change in Cooling Load
1	Increased by 1.12 times
2	Increased by 1.05 times
3	Increased by 1.36 times
4	Increases by 1.23 times
5	Increased by 1.22 times
6	Decreases by 0.98 times

Figure 6 shows heating loads before and after retrofitting for a year as obtained from TRNSYS Studio for the six zones and Table 8 shows changes in heating loads in different zones before and after retrofitting. Heating load decreased approximately by 0.81 in the attic which was the minimum decrease and it decreased approximately by 0.44 times in zone 2. The overall the heating load decreased by 0.67 times when all zones were considered.

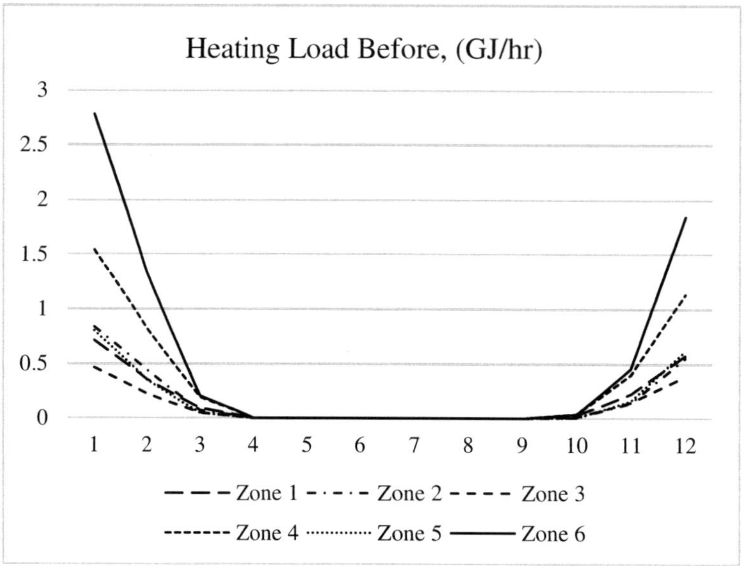

(a)

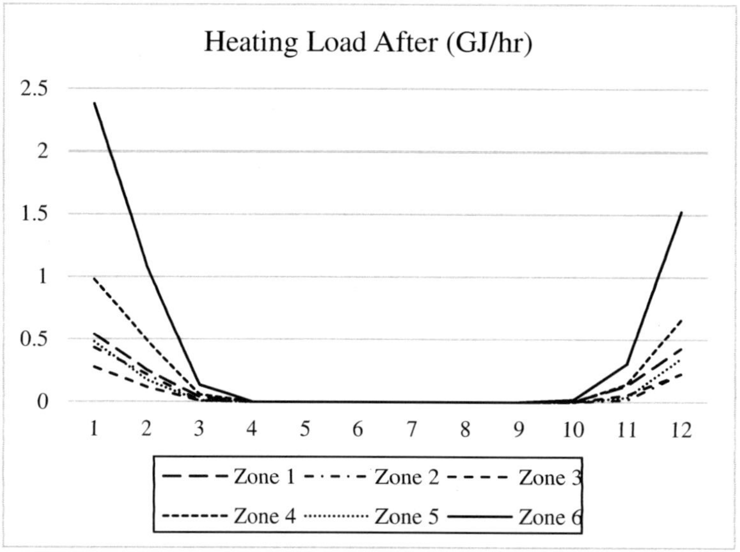

(b)

Figure 6. (a) Monthly heating load before retrofitting (b) Monthly heating load after retrofitting

Table 8. Change in heating load after retrofitting.

Zone	Change in Heating Load
1	Decreased by 0.70 times
2	Decreased by 0.44 times
3	Decreased by 0.54 times
4	Decreased by 0.56 times
5	Decreased by 0.52 times
6	Decreased by 0.81 times

5. Conclusion

The main goal of the study was to assess passive retrofitting strategies that could be used in earthquake-resistant houses to improve the thermal comfort of the residents. This study first selected sites based on two phases of the survey. Thakle Namuna Basti passed through the first phase survey which was visited for the second phase survey. The input for TRNSYS simulation was gathered through the second phase survey. Two cases, with and without passive thermal retrofitting techniques, were compared after simulation. The results showed that the cooling load decreased the most in the kitchen and increased in the attic after retrofitting. Similarly, the heating load decreased the most in the kitchen and increased the most in the attic after retrofitting. This could be due to more heat-generating activities in the kitchen and the use of zinc sheets in the roofing, respectively. The building envelope, windows, and flooring insulation were found to have the potential to reduce heat loss. The inclusion of infiltration in the building and the influence of radiation were not considered in the study but would make the findings more accurate.

6. References

[1] Rafferty J P 2015 *Nepal earthquake of 2015. Encyclopedia Britannica*
[2] Government of Nepal National Planning Commission 2015 *Nepal earthquake 2015 post disaster needs assessment vol. B: sector reports* **8** 67–5
[3] Government of Nepal Department of Urban Development and Building Construction 2015 *Guidelines for earthquake resistant building construction low strength masonry* **203**
[4] Government of Nepal National Building Code 2020 *NBC 105: 2020 seismic design of buildings in Nepal*
[5] Bodach S, Lang W and Hamhaber J 2014 *Climate Responsive Building Design Strategies of Vernacular Architecture in Nepal. Energy and Buildings* **81** 242–227
[6] Rizzo G, Beccali M and Nucara A 2004 *Thermal Comfort. Encyclopedia of Energy* 64–55
[7] Ibáñez M, Lázaro A, Zalba B and Cabeza L 2005 *An approach to the simulation of PCMs in building applications using TRNSYS. Applied Thermal Engineering* **25** 1807–1796

Acknowledgment

The authors would like to acknowledge EnergizeNepal Program, School of Engineering, Kathmandu University, and University Grant Commission, Nepal for their support in carrying out this research.

AUTHOR INDEX

Institute of Physics
Dirac House, Temple Back
Bristol BS1 6BE UK

ISSN: 1757-8981
ISBN 978-1-7138-7142-2